Science and Geopolitics of The White World

Prem Shankar Goel · Rasik Ravindra
Sulagna Chattopadhyay
Editors

Science and Geopolitics of The White World

Arctic-Antarctic-Himalaya

Springer

Editors
Prem Shankar Goel
NIAS
Bangalore
India

and

ISRO
Bangalore
India

Rasik Ravindra
Formerly with Ministry of Earth Sciences (MoES)
New Delhi
India

Sulagna Chattopadhyay
LIGHTS Research Foundation
New Delhi
India

ISBN 978-3-319-57764-7 ISBN 978-3-319-57765-4 (eBook)
DOI 10.1007/978-3-319-57765-4

Library of Congress Control Number: 2017938126

Printed on acid-free paper

This Springer imprint is published by Springer Nature
The registered company is Springer International Publishing AG
The registered company address is: Gewerbestrasse 11, 6330 Cham, Switzerland

Preface

The polar regions, once considered terra incognita due to their inaccessibility and lack of geographical attributes, are known in a much better way now—thanks to the spirit of man's adventurism, quest for scientific knowledge and dawn of the satellite era. Though Arctic was known to Greeks from ancient times and some works as early as 1885 and 1903 (Willian Warren[1]; Bal Gangadhar Tilak[2]) described this region as cradle of human race even before last Glacial Maxima (10,000 BP), the scientific exploration gained seriousness with International Geophysical Year in 1957–58, when a concerted effort was made by the global community to understand several scientific mysteries associated with these areas, especially Antarctica. The concept of Himalaya, as a Third Pole, is rather a recent one.

Considerable progress in the field of science in Arctic and Antarctica has since been made. The science here, however, is intricately linked with geopolitics because of the inherent environmental and strategic interests that transcend the sovereign control of several nations of these regions. With thawing of sea ice in Arctic and strains in Antarctic Treaty, the geopolitics angle is becoming more and more relevant. India has been one of the pioneering members of the Antarctic Treaty System with Consultative Status since 1983. During the period of more than three decades since then, India has added two permanent research stations—*Maitri* and *Bharati*—and contributed significantly to scientific research in earth, atmosphere and biological sciences. As the effectiveness of the Antarctic Treaty is based on the suspension of the territorial claims of claimant nations for the period of the life of the Treaty (*article XII of Antarctic Treaty only talks of 'Review' and not expiry of the Treaty*) and a resolve for 'comprehensive protection of the Antarctic environment and dependent and associated ecosystems' (*Madrid Protocol*), Antarctica is rightly designated as a 'natural reserve, devoted to peace and science'. Even if the Treaty comes up for a review after 2048, it is hoped that the

[1]William F. Warren (1885). Paradise Found or the Cradle of the Human Race at the North Pole.

[2]Bal Gangadhar Tilak (1903). The Arctic Home in the Vedas. Tilak Brothers, Gaikwarwada, Poona City.

Consultative Parties will not like to disturb the status quo so far as mineral exploitation is concerned. Declaration of nearly 1,548,813 km^2 of the sea area, as Ross Sea Marine Reserve in the Antarctic Circle recently, is a step in this direction. It is essential that we carry on this legacy for future generations. Thus, despite the fact that science is the ‘only currency in Antarctica’, perhaps role of the geopolitics is more important here. India should therefore be a proactive player in this field, too.

The Arctic presents a more alarming scenario due to excessive melting of ice and consequent opening of the new areas being thrown open for offshore natural resource exploitation. Most of the area in the Arctic Circle, right up to North Pole, has been claimed by the Arctic nations as their extended continental shelf. Though many of the claims are overlapping, the Arctic states have decided not to contest the conflict before the UN Commission on Limits of Continental Shelf. Moreover, the dwindling sea ice has opened new sea routes in Arctic, making navigation in so far inaccessible area, an easy and economical proposition. This will pave an easy way for bilateral or multilateral cooperation in the resource exploration and profit sharing in the area, posing a serious environmental situation.

Arctic and Antarctic are gaining international limelight in the field of tourism. This might seem contradictory, as on the one hand we are talking of preserving the delicate ecosystem of these pristine areas while on the other hand nobody wants to put a stop to visits of tourists. Since tourism has come to stay as a tool of education and raising environmental awareness, and also because many advanced nations are promoting their tourism industries, the 30th ATCM held in New Delhi in 2007 had suggested a ‘Regulated Tourism’ policy to be adopted wherein number of tourists and their ships could be kept under control. Fortunately, tourism in Antarctica is now very well controlled with an eye on avoiding the introduction of non-native species in its environment.

The three polar regions of Earth—Arctic, Antarctic and Himalaya (as the Third Pole)—also raise an important viewpoint of competitive nationalism vis-a-vis cooperative internationalism. The competitive nationalism, for example, can be described as keeping pace with others or even surpassing them in various spheres of national importance such as raising infrastructure for development, which is a healthy competition. The cooperative internationalism, on the other hand, is joining hands with global community in advancing common goals such as in the areas of human well-being and creating environmental reserves. The scope of latter is far beyond the competitive nationalism as it reduces the global burden and brings humanity closer.

The Himalaya—long considered as a crown of India—constitutes an integral part of every aspect of our life and survival—be it history, culture, agriculture, religion or trade and commerce. Apart from India, this mountain belt encompasses areas under sovereign control of Afghanistan, Bhutan, China, Nepal and Pakistan supporting 1.4 billion people in one of the most densely populated regions of the world (including Bangladesh and Myanmar). The Himalayan glaciers are the sources of several perennial rivers and its high orography controls monsoon rainfall, which is the lifeline of agriculture in this part of the world.

Being geologically a part of dynamic earth system, the regions in Himalaya experience maximum natural hazards such as earthquakes, landslides, avalanches, cloud bursts and glacial lake outbursts. There have been huge losses of life and property. Any major earthquake in the modern times in this region, like the one in Nepal recently, may create havoc with huge loss of life and property due to high population density and rapidly developing infrastructure.

The magnitude of the problem and geographical spread in Himalaya is so great that no single stakeholder can do justice to the subject of such great relevance to the indigenous people inhabiting the inaccessible and inhospitable area, cutting across the geographic borders. In no other region are science and geopolitics so closely entwined as in Himalaya. It is therefore of prime importance that the Himalayan States gain from the successful experience of Arctic States and organize themselves such as Arctic Council and come up with a coordinated plan of Himalayan region development.

The present volume, which is a compilation of 13 papers mainly presented at the SaGAA-III (International Seminar on Science and Geopolitics of Arctic-Antarctic-Himalaya) held at New Delhi on 29–30 September 2015, was a continued attempt, third in succession, by LIGHTS Research Foundation on such a unique subject. The papers cover most of the issues raised in preceding paragraphs and present contemporary analyses of the three polar regions, their scientific issues and geopolitical scenario. We thank the distinguished contributors for their scholarly work.

S/Shri Nilesh Kumar, Ravikant Mahto, Garima Borwankar and Pranav Jain of LIGHTS Research Foundation helped in manuscript formatting. We also express our sincere thanks to Prof. Saraswati Raju of Jawaharlal Nehru University, New Delhi, for scrutiny of the manuscripts.

The editors are grateful to Dr. Satish C. Tripathi, General Secretary, The Society of Earth Scientists, Lucknow, and Honorary Editor, Earth Science India, for critically going through the papers and suggesting modifications in an earlier version of the manuscripts that has greatly improved the quality and content of this volume.

Bangalore, India Prem Shankar Goel

Contents

Part III Global Climate Change and the Himalayan Region

Introduction

The present book 'Science and Geopolitics of the Polar Regions' puts together thirteen selected scholarly papers presented in the Third International Seminar on Science and Geopolitics of Arctic-Antarctic-Himalaya, held at India International Centre, New Delhi, on 29–30 September 2015. The papers represent three main parts of the Seminar and have been grouped accordingly under these three parts that are termed as follows: (a) Geopolitics of the Polar Regions, (b) Global Climate Change and Polar Regions and (c) Global Climate Change and Himalayan Region.

Part I, 'Geopolitics of the Polar Region', touches upon the contemporary views on the international scenario concerning the Arctic and Antarctica. It contains five papers authored by subject experts namely Thorir Ibsen, Alan Hemmings, Bimal Patel, Uttam Sinha and Rasik Ravindra.

Thorir Ibsen, Ambassador of Iceland to India, in his paper 'The Arctic Cooperation—A Model for the Himalayas—Third Pole?', has discussed the structure and evolution of Arctic cooperation and opined that the concept of Arctic Council might be a good model to follow for the Himalayan region, especially in the domain of cross-border scientific cooperation. Interestingly, the International Arctic Science Committee was the forerunner of Arctic Council.

In his paper titled 'The Hollowing of Antarctic Governance', Allen Hemmings, Professor at Canterbury University, Christ Church, New Zealand, and a noted specialist on polar issues, has touched the apparent disabling substantive functions of the Antarctic Treaty System since the time of adoption of Madrid Protocol and the changes that operation of the Antarctic Consultative Committee Meetings (ATCMs) have seen since then. He has discussed the internal inconsistencies due to ever-increasing new instruments added to Antarctic Treaty System, complexities and regulatory gaps.

Dr. Bimal Patel, Professor of International law at the Gujarat National Law University, has deliberated in detail on the state practices on Antarctica and the international law. He comments on the growing influence of new Consultative State Parties, especially the Asian entrants such as China, India and South Korea, their growing involvements and influence on the treaty regime. Prof. Patel has identified and analysed Indian needs and interests and commented on the future perspective.

In the paper 'Arctic: A Paradox and Antithesis', Dr. Uttam Sinha, Adjunct Professor at Malviya Centre for Peace Research, BHU, Varanasi, has emphasized on the need to draw a balance between the predicted change in sea ice and sustainable development of the Arctic Region so that the delicate environment of Arctic Region is protected. He has suggested the need to formulate some polar codes for science and exploration, as in the Antarctic Treaty.

Antarctic tourism has been a subject of nearly all ATCMs in recent years. Dr. Rasik Ravindra, former Chair, Panikkar Professor with Ministry of Earth Sciences, has evaluated the growth and potential of Antarctic tourism and commented on the need for India to encourage regulated tourism to raise the educational value and awareness of environmental issues pertaining to the icy continent.

Part II deals with 'The Global Climate Change and Polar Region'—a topic of intense interest among common men and policy makers alike, as the cryosphere, especially the polar regions are supposed to be the best thermometers for assessing the impact of global warming. This part has two papers, contributed by Dr. M.N. Rajeevan and Dr. S. Rajan.

Dr. Rajeevan, a noted climate specialist and Secretary to the Government of India, Ministry of Earth Sciences, has dealt the subject of linkages between Arctic sea ice phenomenon and the Indian monsoons in his paper titled 'The Arctic—Teleconnections'. He has demonstrated that Arctic oscillation and Indian monsoons are interlinked and sea ice decline has implications of large-scale circulation in mid-latitude. He has further opined that abnormal convection over north-west India could influence the Arctic sea ice.

In his paper, 'Abrupt Climate Shift over the Past 10,000+ Years: An Arctic-Antarctic-Asian Imbroglio?', Dr. S. Rajan, former Director of National Centre for Antarctic and Ocean Research (NCAOR), has commented on the existence of large-scale variability in the monsoons during the last glacial period and the Holocene, based on the records of marine and terrestrial proxies. He has summarized our current understanding of the abrupt climate changes over the past 10,000+ years, focusing on the proxy records of monsoon variability, and their probable linkages with correlative climate change events of the polar regions. He focuses on the frequently asked questions, i.e. whether the short-term linkages between Indian summer monsoons and polar region persisted during past and if so, what was the forcing mechanism?

The third and the final parts on 'The Global Climate Change and the Himalayan Region' contain six papers written by active workers in the field of Himalayan cryosphere. The authors include Dr. Ajai, formerly with Space Applications Centre (ISRO), and currently, CSIR Emeritus Scientist, Dr. M.R. Bhutiyani, Director, Defence Terrain Research laboratory; Dr. H.S. Negi, M.S. Shekhar, H.S. Gusain; and A. Ganju, senior scientists and director, respectively, with Snow Avalanche Study Establishment DRDO; Dr. D. Kumar, A. Choudhary and A.P. Dimri, senior researchers from Jawaharlal Nehru University, Dr. Abul Amir Khan, N.C. Pant from Delhi University and Dr. A.V. Kulkarni, Distinguished Visiting Scientist and Pratibha S. from Divecha Centre for Climate Change, Indian Institute of Science. The part covers a wide spectrum of papers giving inventory, status and fluctuation of the glaciers in Himalaya. The part also deals with papers on regional climate changes, climate variability and mitigation strategies in Himalaya.

Dr. Ajai's paper titled 'Inventory and Monitoring of Snow and Glaciers of the Himalaya Using Space Data' gives details of the recent glacier inventory prepared by SAC, ISRO, that lists 32,392 glaciers on 1:50,000 scale covering an area of 71,182.08 km^2 in Indus, Ganga and Brahmaputra—the three main basins. The paper also brings forth the different rates of retreat of glaciers for the period 1989–90 to 2001–2004 and that for period 2001–02 to 2010–11.

The paper 'Current Status of Himalayan Cryosphere and Adjacent Mountains', by Dr. Abul Amir Khan, N.C. Pant and Rasik Ravindra, points out the contrasting number of glaciers in different inventories and wide variability/uncertainty in data on most of the aspects in the literature on Himalayan cryosphere. The available estimates on snow cover and rainfall, for example, are highly variable. Authors also report a highly heterogeneous behaviour of glacier melting in Hindu Kush Himalaya, e.g. the highest negative mass balance in Eastern Himalaya, relatively less negative in Western Himalaya and positive mass balance in the Karakoram.

Kulkarni and Pratibha in their paper on 'Assessment of Glacier Fluctuations in the Himalaya' have attempted to reconcile the data on glacier fluctuation collected from the ground surveys and remote sensing techniques and have come to a conclusion that the data suggest an overall loss in the glacier area, except in Karakoram region. However, they also point out that there is an apparent discrepancy between different methodologies.

Climate changes impact glacier health considerably. Kumar and his co-workers (A. Choudhary and A.P. Dimri) in their paper 'Regional Climate Changes Over Hindukush-Karakoram-Himalaya Region' have shown that in spite of scanty in nature, the model dataset shows a significant warming trend in temperature as also a positive trend in both summer and winter monsoon precipitation over western Himalaya, consequently impacting the glaciers.

The research group comprising H.S. Negi, M.S. Shekhar, H.S. Gusain and A. Ganju from SASE in their paper on 'Winter Climate and Snow Cover Variability Over North-West Himalaya' have analysed the trends of climate and snow cover in three different climatic zones of north-western Himalaya viz. Lower, Great and Karakoram Himalayas. Their analysis of past two decades suggests that the maximum temperatures have increased and minimum temperatures have decreased in all the three zones. The trends of winter mean temperatures have been found increasing in Lower and Great Himalayas and decreasing in Karakoram Himalaya.

M.R. Bhutiyani in his paper titled 'Mitigation Strategies to Combat Climate Change in the Himalayan Mountains' has dealt with effects of melting of glaciers on the hydro-meteorological regimes of river basins in the Himalaya and suggested the use of innovative environment-friendly technologies to tap renewable sources of energy such as solar, wind, hydro-electric and geothermal energy.

Prem Shankar Goel
Rasik Ravindra
Sulagna Chattopadhyay

Part I
Geopolitics of the Polar Region

The Arctic Cooperation, a Model for the Himalayas—Third Pole?

Thórir Ibsen

Abstract The challenges faced by the peoples and governments in the Himalayas-Third Pole region resemble the ones in the Arctic. Both regions are seriously affected by external pressures such as climate change, and they share similar geopolitical circumstances. What differs between the two is that one has an effective regional governance structure to address these challenges while the other does not. The question therefore arises whether one could learn from the other. The simple answer to the question is yes. The experience from the bottom-up and science based evolution of the Arctic cooperation could serve as a model to build trust, bridge tension and foster intergovernmental cooperation that deals specifically with the Himalayan-Third Pole region. However, the politics of the region are complex and highly influenced by security tensions. It would appear that active leadership of the dominant powers, India and China, would be necessary. These two have the capacity and the resources, and as observer states to the Arctic Council, they have the first hand insight and experience of the Arctic governance structure.

Keywords Himalayas-Third Pole · Arctic · Geopolitics · Climate change · India

1 Introduction

There is substantial scholarly interest in Polar research in India. Some is directed to pure science and other to governance in the Arctic and the Antarctic. This can be witnessed by the numerous Polar conferences held in India and by an impressive list

---

Ambassador of Iceland to India. The views expressed in this article are those of the author and do not represent the views of the Government of Iceland.

T. Ibsen (✉)
Embassy of Iceland, New Delhi, India
e-mail: thorir.ibsen@utn.stjr.is

P.S. Goel et al. (eds.), *Science and Geopolitics of The White World*,
DOI 10.1007/978-3-319-57765-4_1

of publications.[1] Historically there have been different motives behind this scholarly work, extending from searching ancestral origins in the North,[2] through interest in scientific research, to a sense of need to defend the global position of India in world politics (Sinha and Gupta 2014; Chaturvedi 2013a, b, c, 2014).

India became engaged in the Antarctic in the early 1980s, and with the emergence and development of the Arctic cooperation, India like many Asian countries has deemed it necessary to be engaged in this Northern region. India is one of the five Asian countries that have obtained observer status in the Arctic Council. The other four are China, Japan, South Korea and Singapore. The participation of these countries is welcomed by the member states of the Arctic Council. They view their involvement as important for strengthening scientific research in the region and for maintaining the Arctic on the global agenda.

The motives of the five Asian countries are at once similar and different. Each one of these countries has its own specific reasons for participating in the Arctic cooperation, likewise their areas of engagement are different and their level of participation varies. Yet one can discern from observation that all are driven by three common, albeit not necessarily consistent motives: (1) the benefits of research cooperation, (2) a sense of global responsibility to protect the Arctic as "global commons" or "common heritage of the mankind", and (3) an urge to defend their national interests in future economic activities in the Arctic which extend from exploitation of natural resources to trans-polar shipping.[3]

In this complex constellation of altruism and interests of the Asian countries, it is also occurring that lessons can be learned from the Arctic cooperation to advance peace, security and prosperity in other regions. This is in particular true for the Himalayan region where smaller populations and nations live in the shadow of tensions between large powers while also experiencing increased social and

[1]A good example is the "SaGAA" conference series of the LIGHTS Research Foundation of 2011, 2012 and 2015 which has brought together leading Indian and international researchers and scientists in the area of Polar research. Many of the papers are available in this volume and in the previous publication edited by Ramesh et. al (2013). See also IDSA (2014) which contains papers presented at the conference "Geopolitics of the Arctic: Commerce, Governance and Policy" held at the Institute for Defence Studies and Analyses (IDSA) in New Delhi in September 2013.

[2]Dr. Sanjay Chaturvedi, Professor at the Department of Political Science at Panjab University drew the attention of the author to the book of Bal Gangadhar Tilak *The Arctic Home in the Vedas* in which Tilak establishes ancestral connection of India to the Arctic arguing that the *Vedas* were composed in the Arctic and that the Aryans had brought them south when they fled the last ice age. What is interesting is that this book was written in 1898 (first published in 1903) by an Indian nationalist whom the British colonial authorities are said to have called the "father of the Indian unrest" while his followers gave him the honorary title "Lokmanya", or the one "accepted by the people (as their leader)". https://en.wikipedia.org/wiki/The_Arctic_Home_in_the_Vedas; https://en.wikipedia.org/wiki/Bal_Gangadhar_Tilak.

[3]These motives were recurrent themes in presentations of researchers from the five Asian states at the *Annual Maritime Power Conference of the National Maritime Foundation devoted to the topic Asia and the Arctic: Opportunities and Challenges,* held in New Delhi 19–20 February 2015. See also IDSA (2014) and Roundtable (2014).

economic vulnerability caused by climate change. It is less clear, however, whether the larger players in the region do see their interest in bringing the practical experience of the Arctic cooperation to bear on the Himalayan region.

There are indeed number of resemblances between the Arctic and the Himalayas, which is often referred to as the Third Pole having the third largest ice cover in the world. Notably both regions are facing challenges of anthropogenic causes where climate change is a major contributor. Both regions possess pristine natural environment and biological diversity that must be conserved. They both have populations that draw their livelihood from these local resources and both regions have communities that are dealing with rapid social and economic changes. Both regions have a large ice-cover of significant importance for the environment, biological diversity, economic activities and for water management. Lastly but not the least, both regions have experienced security tensions, militarization and territorial and border disputes. What differs, however, between the two is that while the Arctic states have built a successful governance structure for their cooperation, far less intergovernmental cooperation exists between the Himalayan countries to address the particularities and challenges of the region.

Ólafur Ragnar Grímsson has made a strong case for the relevance of the Arctic cooperation as an inspiration for building the trust that is necessary for systematic intergovernmental cooperation in the Himalayas—Third Pole region.[4] He argues that just as research and science cooperation was a key in bridging between the former adversaries of the Cold War, leading to the formation of the Arctic Council, the same approach could be applied in the Himalayan region. Thus, research cooperation for gathering and sharing scientific data about the common denominator, the ice, its transformation and its implications for the region and beyond, could become the basis for informed diplomatic and political cooperation in the Himalayas and for successful regional policies (Grímsson 2015).

This article explores this proposition by drawing up a picture, albeit not a complete one, of the challenges faced by the two regions in the Arctic and the Himalayas and of how the peoples, public authorities and States in these two regions have sought to respond collectively to these challenges. The article does not purpose to answer all the aspects of the question of whether the Arctic model is feasible or indeed practicable for the peoples and governments of the Himalayan region. Rather, the purpose of the article is more to address the question with a view to stimulating more research and dialogue about future governance in the Himalayas-Third Pole region.

---

[4]Dr. Ólafur Ragnar Grímsson, former President of Iceland and accomplished scholar of political science has devoted much attention to the Arctic and the Himalaya regions, not the least in the context of climate change. He is the founder of the Arctic Circle, an independent forum that brings together in an open dialogue decision-makers, business leaders, scientists and representatives of civil society. The Arctic Circle holds annual Assemblies in Reykjavík. It also organizes smaller forums on specific issues, such as the 2015 forums in Alaska and Singapore and the 2016 forums in Québec and Greenland. http://arcticcircle.org/about.

2 The Arctic and Its Challenges

What distinguishes the Arctic and the Himalayas-Third Pole from the Antarctic, to state the obvious, is that the two former regions are populated by peoples and nations that have interests and territorial borders. In a word, these peoples and nations rely on economic exchange and have a history of military build-up, tension and border disputes. Their current and future relations and living conditions are to a large extent determined by the large ice cover found in both regions that dictate access to resources, transport-routes, ecosystem services and meteorological conditions. The single most important development affecting the future of the two regions is climate change. It brings both threats and opportunities. For the peoples and the governments, the challenge is to balance these factors and prepare for the opportunities by addressing the risks in a responsible and rational manner. These threats and opportunities are well documented in the Arctic (ACIA 2004; Stefansson Arctic Institute 2004; Larsen and Fondahl 2014).

The melting of the Arctic ice brings both new risks and opportunities. Rich natural resources, including hydrocarbons, minerals and fisheries will be revealed. The changes in the climate may also alter ocean currents and thereby the distribution, even presence of commercial fish stocks. Likewise, growing number of cruise ships in the Arctic Ocean along with the development of new shipping routes between the Pacific and the North-Atlantic through the North-East passage and even over the North Pole will bring new economic opportunities but also new dangers.

With these ample opportunities there are obviously a range of possible future scenarios for the development of the Arctic. The challenge is to ensure that the development be peaceful, prosperous and sustainable and that it does not lead to limitless competition for natural resources that could relapse into re-militarization of the scale experienced during the Cold War era.

The Arctic cooperation emerged largely to deal with the devastating environmental legacy of the Soviet Union (Gordon Foundation 2010; English 2013; Koivurova and Hasanat 2008; Chater 2015). Concerns were also with growing signs of pollution being brought to the Arctic by wind and sea currents from distant places, and there was growing recognition of the need to address the conservation of the vulnerable Arctic biological diversity. Scientific knowledge was required to assess the impact of pollution on the physical environment and on the biological diversity in order to propose and develop plans and programmes for action. Some environmental issues required concerted action among the Arctic States, others required soliciting the support of the global community in stemming the tide of long-range air and water pollution.

As the cooperation evolved, new challenges and concerns emerged. First and foremost were the concerns over the available evidence of the impacts of serious climate change in the Arctic. The second was the growing recognition that the Arctic is not simply a pristine environment but a home to 4 million people that supply food and other vital natural resources to hundreds of million people outside the region. The Arctic cooperation offered also an opportunity to improve living conditions, services and economic activities in the region, promote regional business cooperation and to take charge of the changes taking place as a result of climate change so as to minimize the negative impact and take advantage of new opportunities. These changes in turn have brought in new risks and challenges that can only be addressed through enhanced regional cooperation.

The Arctic States face a number of current and future challenges. To mention a few salient ones, the growing pressure to utilize the resources within the Arctic and the opening of new shipping routes between the Pacific and the North-Atlantic will increase disagreements and disputes on rights to resources, territory and passage. Thus the Arctic States must ensure that the governance structure of the Arctic region, the Arctic Council and the UN Convention on the Law of the Sea (UNCLOS) continue to provide effective and well functioning diplomatic and legal dispute resolution mechanism for the region. Likewise the Arctic states must provide for effective surveillance cooperation along with search and rescue capacity to respond to fatal sea accidents or marine-pollution. There are growing risks stemming from increasing sea-traffic of huge cruise ships with massive numbers of tourists and there are risks of oil spills caused by accidents that can happen during exploitation and transportation of natural resources in the North. Moreover, a necessary infrastructure is required to deal with the growing trade, tourism and transportation and to support sustainable economic cooperation in the region. Lastly, there is further need for strong research cooperation to ensure sound scientific knowledge and effective monitoring of environmental changes in the region.

3 Lessons from the Arctic Cooperation

Looking to the origins of the Arctic cooperation, it has to be kept in mind that after the WW II the Arctic region became heavily militarized—in fact the Arctic became a frontier in the Cold War. With the end of the Cold War in the 1990s, a political space opened for constructive cooperation. Old foes recognized the need to work together in mutually beneficial ways to deal with common threats stemming from trans-boundary pollution and climate change. It started with scientific cooperation leading to the establishment of the International Arctic Science Committee (IASC) in 1990, a non-governmental international scientific organization; followed by the Northern Forum created in 1991 by local and regional governments; and lastly the Arctic Environmental Protection Strategy (AEPS) launched by the 1991 Rovaniemi Declaration of all the eight Arctic states: Canada, Denmark, Finland, Iceland, Norway, Sweden, the USSR and the United States.

The AEPS was an initiative of the Environment Ministers of the Arctic countries. It entailed four programmes focused on environmental assessment, monitoring, emergency response and conservation, namely the Arctic Monitoring and Assessment Programme (AMAP), Emergency Prevention, Preparedness and Response (EPPR), Protection of the Arctic Marine Environment (PAME), and

Conservation of Arctic Flora and Fauna (CAFF). The four programmes were run by four working groups of experts from the Member States and supported by secretariats located in Norway and Iceland. The overview of the implementation of the AEPS was in the hands of senior civil servants, entitled Senior Arctic Officials (SAOs). The SAOs prepared the meetings of Environment Ministers and supervised the implementation of their decisions and recommendations. A key stake-holder, the indigenous peoples' organizations did not have a formal status but attended meetings and lobbied state representatives, receiving support from the Indigenous Peoples Secretariat initially hosted by Denmark, but now located in Tromsø, Norway.

As the awareness and recognition of the importance of the Arctic grew, not only for environmental protection but also for human development, economic activity and for security, a dialogue evolved among the member states to raise the cooperation to another level, leading to the establishment of the Arctic Council in 1996.[5] The Council's mandate was to promote cooperation on sustainable development in the Arctic region as well as environmental cooperation. The Arctic Council inherited the organizational structure of the AEPS but now under the leadership of Foreign Ministers instead of Environment Ministers. Also a new working group was established in 1996 to address the human dimension of the Arctic, the Sustainable Development Working Group (SDWG), with secretariat support in Canada.

The Arctic Council was established as a forum for political consultation and policy making of Foreign Ministers supported by a permanent body of Senior Arctic Officials (SAOs), six Working Groups and number of expert groups and task forces.[6] A novelty in the set-up of the Arctic Council is that the indigenous peoples' organizations do not only have an observer status at the meetings of Ministers, the SAOs and the working groups, they have a seat at the table as Permanent Participants.[7] While they do not have the full rights of the Member States, they can participate in the meetings, the decision-shaping, agenda setting and in initialising projects. The Arctic Council also involves over 32 observer states, international organisations and NGOs.[8]

The Arctic Council was formed as a project based intergovernmental forum, but it has evolved towards assuming more policy related responsibilities and permanent organizational structure. When created, there was no political will to establish a

[5]The Arctic Council was founded by all the eight Arctic states, by this time the Russian Federation had replaced the USSR.

[6]The sixth working group, the Arctic Contaminants Action Program (ACAP) was added in 2006. http://www.arctic-council.org/index.php/en/about-us.

[7]Six organizations representing Arctic Indigenous peoples have status as Permanent Participants: the Aleut International Association, the Arctic Athabaskan Council, Gwich'in Council International, the Inuit Circumpolar Council, Russian Association of Indigenous Peoples of the North and the Saami Council. http://www.arctic-council.org/index.php/en/about-us/permanent-participants.

[8]http://www.arctic-council.org/index.php/en/about-us/arctic-council/observers.

permanent secretariat even though the working groups had their own permanent secretariats. Thus each chairmanship assumed the responsibility to operate a small secretariat for the duration of its two-year term. Finally in 2011 an agreement was reached to establish a permanent secretariat in Tromsø, Norway.

The Arctic Council neither has a common budget nor a legal status. The secretariats and individual projects are funded by host countries and through voluntary contributions of Member States. The Arctic Council is not based on an international treaty with the statute of international law like is the case with the Antarctic Treaty. As such it does not entail legal commitments or a legally binding dispute resolutions mechanism. This fact has led to the recurrent discussion of whether there is a need for developing a specific legal regime for the Arctic, similar to the Antarctic Treaty System. [9] However, as the Arctic is mainly an oceanic area Member States have been of the view that there is no need for such a treaty in the light of the existence of the Law of the Sea Convention (UNCLOS) which covers sufficiently the rights and obligations of states in the seas of the Arctic and provides for an effective dispute resolution mechanism.

Yet the political scope and policy-making importance of the Arctic Council is growing. Thus the human dimension of the Arctic, sustainable development and the implications of climate change are now recognized as equally important to the initial environmental protection agenda of the AEPS. Two major reports published in 2004, the *Arctic Climate Impact Assessment* (ACIA) and the *Arctic Human Development Report*, played an important role in that transformation. Also the Council has grown from promoting primarily cooperation on scientific research and providing recommendations to actually taking decisions that impose obligations on Member States with the adoption of two legally binding agreements, the *Agreement on Cooperation on Aeronautical and Maritime Search and Rescue in the Arctic* of 2011, that came into force in 2013 and the *Agreement on Cooperation on Marine Oil Pollution, Preparedness and Response in the Arctic* of 2013.[10]

There are also a number of parallel forums that are important for fostering cooperation in the region and for providing support for the Arctic Council. These include the Conference of Parliamentarians of the Arctic Region, the Arctic Economic Council, the World Winter Cities Association for Mayors, and the Youth Arctic Coalition. The latest addition is the Arctic Circle, an independent forum established in 2013. While not attached to the formal structure, the Arctic Circle has the additional value of being more flexible and inclusive than other forums in the region, as it is neither intergovernmental nor formed around particular participants. The Arctic Circle works on broadening and deepening the dialogue on the Arctic by bringing to the table more actors from different walks of life, notably from civil society, the business sector, the scientific community as well as public policy makers, both from within and outside the region.

---

[9]The Antarctic Treaty of 1961 and related agreements are collectively known as the Antarctic Treaty System (ATS).https://en.wikipedia.org/wiki/Antarctic_Treaty_System .

[10]http://www.arctic-council.org/index.php/en/our-work/agreements.

The benefits and utilities of the Arctic Council in particular and the broader Arctic cooperation in general have many dimensions. The Arctic States have the mechanism to collectively identify, address and resolve shared problems and common concerns in the region. They have the means to pull their resources and produce more comprehensive and credible reports on the state of the Arctic region and its socio-economic development. They have the political venue to coordinate policies and for calling up on the global community to help address external threats to the region. The cooperation has raised public awareness, both within and outside the region about the Arctic and its importance and it provides an effective venue for the indigenous peoples to raise their concerns and pursue their specific interests in the region and beyond. Moreover, the cooperation has strengthened national institutions and agencies that are tasked with Arctic research, science, monitoring and surveillance as well as the development and implementation of domestic policies and programmes. Lastly, but not the least, the cooperation allows scientists, institutions, agencies, NGOs, the indigenous peoples, the business community, local authorities and policy-makers to raise their particular concerns and to build cooperation with partners within the region and beyond.

4 The Himalayas—The Third Pole

Owing to the presence of a vast number of glaciers, the Himalayas are often referred to as the Third Pole. This appellation signifies its semblance to the two Poles, the Arctic and the Antarctic. It also recalls the importance of the three ice covered regions for global meteorological conditions and their shared vulnerability to climate change.

A quick glance reveals the geographical, environmental and political significance of the Himalayas-Third Pole region. If the region is defined narrowly meaning the imposing Himalayan mountain range,[11] it links five countries: Pakistan, India, Nepal, Bhutan and China. This mountain range has the third largest deposit of ice and snow in the world. Only the Antarctic and the Arctic have more. From the glaciers of the Himalayas run three of the world's major rivers—the Indus, the Ganges and the Brahmaputra. Their combined drainage basin is home to some 600 million people.

If defined more broadly to cover the extended Himalayan region including the Hindu Kush mountain range and the Tibetan Plateau range,[12] the Third Pole region spans an area of more than 4.3 million km^2 and includes 9 countries, that is Afghanistan, Tajikistan, Bangladesh and Myanmar in addition to the above mentioned five countries. Some 210 million people live in the region, composed of different ethnic communities speaking 600 languages and many more dialects. This

[11] https://en.wikipedia.org/wiki/Himalayas.

[12] http://www.icimod.org; http://www.thethirdpole.net/about.

larger Third Pole region has even broader implications as it is the source of 10 major river systems[13] providing irrigation, power and drinking water for over 1.3 billion people in Asia, or close to one fifth of the world's population. More than 3 billion people benefit from the agricultural products and energy produced in these river basins. To this can be added that the region is a highly attractive tourist destination and has spiritual significance for hundreds of millions of Buddhists, Hindus, Jain, Sikhs and Muslims.

Climate change is a major concern for the region. The melting of glaciers and the snow cover impacts meteorological conditions, rivers and water reservoirs and thereby alters water management, electricity production and agriculture and thus the food security and economic well-being of millions of people. It also leads to exposure of new lands and possibly new natural resources, and causes changes in the landscape and the structure, composition and distribution of the unique Himalayan flora and fauna. The implications go far beyond the Himalayan region, as its agricultural production supplies distant markets with vital agricultural products and rivers from the region provide water, irrigation and electricity in distant places. Moreover, accelerated melting of the ice-cover affects global meteorological conditions and ecological conditions in distant seas like the Arabian Sea and the Bay of Bengal.

As in the Arctic, the changes that are taking place in the Himalayas will continue, possibly at a more accelerated pace. The speed of the changes might be slowed down by global mitigation efforts but they cannot be halted. Thus for the people living in the region the immediate question is how to adapt to these changes over the next few decades. For policy makers the key questions are how to predict and manage these changes and transformations in a peaceful and constructive manner. These questions in turn call for informed decisions based on thorough scientific knowledge and monitoring of the state and changes of the Himalayan glaciers and effective information exchange and scientific cooperation in the region.

5 Regional Cooperation in the Himalayas

The Himalayas—Third Pole region is served with less intergovernmental cooperation than the Arctic region whether in the area of scientific research or policy formulation. Substantive scientific competence and capacity exists in the region but it appears that systematic information exchange and scientific cooperation across borders are limited. Moreover, while intergovernmental frameworks exist in the region, they seem to lack either political focus on the Himalayas or they do not have the political weight to address the challenges in the region.

On the scientific front, both China and India have substantial competences. The largest Himalayas related scientific programme in China is without doubt the

[13]The Amu Darya, Indus, Ganges, Brahmaputra (Yarlungtsanpo), Irrawaddy, Salween (Nu), Mekong (Lancang), Yangtse (Jinsha), Yellow River (Huanghe), and Tarim (Dayan).

Institute of Tibetan Plateau Research, co-founded by the Chinese Academy of Sciences and the Max-Plank Institute in 2003. It has three campuses in Lhasa, Beijing and Kunming and operates the Tibetan Observation and Research Platform (TORP) that comprises five field stations on the Tibetan Plateau.[14]

In India, glaciological studies in the Himalayas are undertaken by a number of institutes including the Geological Survey of India, the Space Application Centre, the Wadia Institute of Himalayan Geology, and the Divecha Centre for Climate Change (Ravindra and Laluraj 2012). Furthermore, in an effort to step up research, information exchange and policy response related to the implications of climate change in the Himalayas, the Government of India has established a National Mission for Sustaining the Himalayan Ecosystem (NMSHE). The Mission is one of eight national missions launched by the National Action Plan on Climate Change (NAPCC) (Govt. of India 2010, 2014; Venkataramani 2015).

Both India and China have also through their research activities in the Antarctic and the Arctic, built up substantial scientific competence and research capability of relevance for scientific research cooperation in the Himalayas. The Indian Antarctic Program and research station, led by the National Centre for Antarctic and Ocean Research (NCAOR) dates back to the early 1980s and the more recent Himadri station located on the Arctic island of Svalbard was set up in 2008.[15] The Chinese Arctic and Antarctic Administration (CAA) operates two research stations in the Antarctic built in the 1980s and is starting the third station, and it runs a large scientific research base on Svalbard Island opened in 2004.[16]

Looking towards the intergovernmental framework, the principal regional organization is the South Asian Association for Regional Cooperation (SAARC). While recognising the importance of the Himalayas, its glaciers and the threat of climate change, it seems that the organization does not support programmes specific for the Himalayan region such as systematic research cooperation. SAARC's focus is on conventional economic cooperation of the South Asia region. The membership is also at once wider and narrower than the Himalayan region; there are non-Himalayan members and there is an important Himalayan country that is an observer.[17]

A more specialized regional organisation is the International Centre for Integrated Mountain Development (ICIMOD), an autonomous international institution situated in Kathmandu in Nepal.[18] Established in the early 1990s with the backing of UNESCO and support of Germany and Switzerland, ICIMOD is a learning and knowledge centre serving the eight regional member countries of the

---

[14]http://english.itpcas.cas.cn.

[15]https://en.wikipedia.org/wiki/Indian_Antarctic_Program; http://www.ncaor.gov.in; https://en.wikipedia.org/wiki/Himadri_Station.

[16]https://en.wikipedia.org/wiki/Chinese_Arctic_and_Antarctic_Administration; http://www.chinare.gov.cn/en/; http://www.china.org.cn/english/features/PolarResearch/168048.htm.

[17]http://www.saarc-sec.org.

[18]http://www.icimod.org.

Hindu Kush Himalayas—Afghanistan, Bangladesh, Bhutan, China, India, Myanmar, Nepal, and Pakistan. The centre is devoted to improving living standards in the extended Himalayan region and assists mountain people in understanding and adapting to the changes taking place as a result of climate change. ICIMOD works with the United Nations Environment Programme (UNEP)/GRID-Arendal and the Center for International Climate and Environmental Research—Oslo (CICERO) on developing the Himalayan Climate Change Adaptation Program (HICAP). The project, launched with financial support from Norway, engages leading institutes from the region (Das 2015).

ICIMOD is thus engaged in research and survey of glaciers across the Himalayas, supports human development in the region, and facilitates networking, exchange of knowledge and experience. It also seeks to offer a regional platform where policy makers, experts, planners, and practitioners can exchange ideas and perspectives towards the achievement of sustainable mountain development. However, ICIMOD, appears to function primarily with the support of resources coming from outside the region, and as its activities are strictly of scientific and technical nature it does not have a high-level political governance structure.

The relative absence of systematic cross-border scientific cooperation in the region and lack of effective regional governance is doubtless the consequence of the tension and border disputes between China, India and Pakistan. Indeed the Himalayas are a highly militarized security area. A mere daily observation of the Indian media suffices to appreciate the extensive militarization and security arrangements in the region. It is thus not unreasonable to conclude that security and defence are the principal policy-focus of these three States in the region, and that this policy focus overshadows the possibilities for cross-border cooperation in the area of scientific research.

Interest in establishing more systematic intergovernmental governance for the region is nonetheless there. An interesting platform has emerged inspired by the Arctic Circle. This is the Himalayas-Third Pole Circle, of which first formal meeting was hosted by the Government of Bhutan in Thimphu on February 5–6, 2015.[19]

The Himalaya-Third Pole Circle has evolved through a series of meetings organised by ICIMOD, the Third Pole Environment (TPE) programme[20] and the Arctic Circle (Grímsson 2015; Sveinbjörnsson 2014). All three seek to promote regional and international cooperation in the area of science, research and sharing of information with a view to encouraging cross-border science-policy dialogue and informed decisions on adaptation to changes brought about by climate change. The

---

[19]http://www.tibet328.cn/11/2014/07/9/201408/t20140808_439028.htm; http://www.tpe.ac.cn/node/220; http://www.gwp.org/en/gwp-south-asia/GWP-SAS-IN-ACTION/News-and-Activities/Third-Pole-Meeting-in-Thimpu-Bhutan; http://www.icimod.org/?q=17399; http://www.bbs.bt/news/?p=48382; http://www.icimod.org/?q=11645.

[20]Located at the Chinese Academy of Science in Beijing, the Third Pole Environment (TPE) programme is the result of the efforts of the Institute of Tibetan Plateau Research to form an international research programme on environmental changes and adaptation in the Third Pole region. http://www.tpe.ac.cn.

concept of the Himalaya-Third Pole Circle is based on the Arctic experience. The point of departure is the bottom-up and science based approach of fostering relationships across national borders between scientific institutions in the region. The sharing of scientific data and research cooperation in turn establishes systematic dialogue between scientists and public and private policy-makers, that can build regional trust and support informed policy response to the challenges faced by peoples and governments in adapting to climate change.

It is of course too early to assess the Himalaya-Third Pole Circle or to predict its future. Its formation is a testimony to the proposition that the Arctic experience can serve as a model for the Himalayan region. It will be interesting to observe how the dominant powers in the region will respond to this open invitation to bring the region together into an effective cooperation.

6 Conclusions

The motive behind this short article was the question of whether the experience of the Arctic cooperation offers a model for building effective intergovernmental governance in the Himalayas-Third Pole region. The simple answer to the question is yes, given the similarities of the challenges in the two regions and their geopolitical circumstances. Both face serious external challenges such as from climate change and both have experienced tension and conflict over resources and borders. What differs between the two is that there is an effective regional governance mechanism in the Arctic while this has not yet been developed for the Himalayas-Third Pole region.

Thus in brief, it can be argued that the experience from the bottom-up and science based evolution of the Arctic cooperation could help in building the trust that is necessary for meaningful intergovernmental cooperation in the Himalayas-Third Pole. More so, the two dominant powers that have the necessary competence and resources, India and China, are among the five Asian countries that have obtained an observer status in the Arctic Council. Apart from benefitting the Arctic cooperation, the engagement of India and China also endows them with rich and pragmatic experience that can help bridge tension, build trust and foster constructive dialogue on the basis of scientific cooperation in the Himalayas-Third Pole region.

The more complex question is whether the political climate is favourable for establishing such a comprehensive intergovernmental cooperation in the Himalayas-Third Pole region. In a way this process has already started, and has received a new impetus with the establishment of the Himalaya-Third Pole Circle. However, these first steps appear not yet to have the same degree of political commitment and level of participation of States as was the case in the Arctic.

One thing is though certain and that is that the geopolitical importance of the Himalayas-Third Pole region will only increase in the coming years, not the least with the rapid climate change. As science has confirmed (IPCC 2014), climate change cannot be halted, thus peoples and nations in the Himalayan region will have to develop strategies to adapt to the accompanying changes. This adaptation can take place through constructive intergovernmental cooperation, or in the context of conflicts with subsequent destructive consequences. The experience of the Arctic cooperation has demonstrated that a meaningful peaceful intergovernmental cooperation is both possible and feasible in regions that have experienced serious political tensions and military build-ups. And what is more, the cooperation benefits all—nations and indigenous peoples alike. But such a cooperation requires political will and commitment, and to be useful it must be open, inclusive and democratic and provide an effective mechanism to deal with disagreements and resolve such disputes that arise from claims to resources, territory or borders.

References

ACIA (2004) Impacts of a warming Arctic: ACIA overview report. Cambridge University Press, Cambridge

Chater A (2015) Explaining the evolution of the Arctic Council. PhD dissertation. University of Western Ontario. http://ir.lib.uwo.ca/etd/3230

Chaturvedi S (2013a) Antarctic as 'Global Knowledge Commons': geopolitics, science and trusteeship. In: Ramesh R, Sudhakar M, Chattopadhyay S (eds) Scientific and geopolitical interests in Arctic and Antarctic. Lights, New Delhi

Chaturvedi S (2013b) Rise and decline of Antarctic in Nehru's geopolitical vision: challenges and opportunities of the 1950s. Polar J 3(2): 301–315

Chaturvedi S (2013c) China and India in the 'Receding' Arctic: rhetoric, routes and resources. Jadavpur J Int Relat 17(1): 41–68

Chaturvedi S (2014) India's Arctic engagement: challenges and opportunities. Asia Policy 18:73–79

Das R (2015). Preserving Hindu Kush eco zone. Gov Today 01 (12): 51–54. http://governancetoday.co.in/preserving-hindu-kush-eco-zone

English J (2013) Ice and water: politics, peoples and the Arctic Council. Allen Lane, Toronto

Gordon Foundation (2010) A brief history of the creation of the Arctic Council. Gordon Foundation, Toronto

Govt. of India (2010) Mission document. Government of India Department of Science & Technology, Ministry of Science & Technology, New Delhi. http://164.100.94.111/sites/default/files/NMSHE_June_2010.pdf

Govt. of India (2014) Approval for the national mission for sustaining Himalayan Ecosystem launched under the national action plan on climate change. Press Information Bureau Government of India, Ministry of Science & Technology. http://pib.nic.in/newsite/PrintRelease.aspx?relid=104353

Grímsson ÓR (2015) A speech by the president of Iceland Ólafur Ragnar Grímsson at the Himalayas-Third-Pole Circle. Thimphu Bhutan. http://english.forseti.is/media/PDF/2015_02_05_Butanraeda.pdf

IDSA (2014) Arctic: commerce, governance and policy. Strateg Anal 38(6)

IPCC (2014) Climate change 2014: synthesis report. In: Pachauri RK Meyer LA (eds) [Core Writing Team] Contribution of working groups I, II and III to the fifth assessment report of the intergovernmental panel on climate change. IPCC Geneva, Switzerland. http://www.ipcc.ch/report/ar5/syr

Koivurova T, Md Hasanat W (2008) The climate policy of the Arctic Council. December 2008, research gate http://www.researchgate.net/publication/226295962

Larsen JN, Fondahl G (eds) (2014) Arctic human development report: regional processes and global linkages. Nordic council of ministers. http://library.arcticportal.org/1840/1/AHDRFULLTEXT01.pdf

Ramesh R, Sudhakar M, Chattopadhyay S (eds) (2013) Scientific and geopolitical interests in Arctic and Antarctic. Lights, New Delhi

Ravindra R, Laluraj CM (2012) Cryosphere research: Indian perspective. Proceedings of the Indian National Science Academy 78(3):249–257. http://www.insa.nic.in/writereaddata/UpLoadedFiles/PINSA/Vol78_2012_3_Art02_249_257.pdf

Roundtable (2014) Polar pursuit: Asia engages the Arctic. Asia Policy 18:1–79

Sinha UK, Gupta A (2014) The Arctic and India: strategic awareness and scientific engagement. Strateg Anal 38(6):872–885

Stefansson Arctic Institute (2004) Arctic human development report. Stefansson Arctic Institute, Akureyri

Sveinbjörnsson D (2014) Himalaya-Third Pole Circle: forum for dialogue built on the Arctic experience. 2014-08-08 Tibet328.cn

Venkataramani V (2015) National mission for sustaining the Himalayan Ecosystem. Centre for Development Finance (CDF), IFMR LEAD, one of the eight-brief series prepared as part of the Study titled 'Implementation of the National Action Plan on Climate Change (NAPCC)—progress & evaluation'

The Hollowing of Antarctic Governance

Alan D. Hemmings

Abstract The paper examines the apparent disabling of some substantive functions of the Antarctic Treaty System (its 'hollowing') since the adoption of the Madrid Protocol in 1991. It provides some examples of such hollowing by reference to regulatory gaps, the Antarctic continental shelf, and changes in the operation of Antarctic Treaty Consultative Meetings, before exploring the drivers of the hollowing and future options to reinvigorate Antarctic governance.

Keywords Antarctic Treaty · Antarctic Treaty System · International relations

1 Introduction

The system of Antarctic governance is now long established and, seemingly, well grounded. The foundational instrument, the Antarctic Treaty[1] was adopted in 1959; it is given effect through now annual diplomatic meetings—the Antarctic Treaty Consultative Meeting—rotating through Consultative Party states; and from this platform a wider Antarctic Treaty System[2] (ATS) has been developed, addressing

[1]The Antarctic Treaty. Adopted Washington DC 1 December 1959, entered into force 23 June 1961. 402 UNTS 71.
[2]The Antarctic Treaty System is defined in Article 1 of the Madrid Protocol (below) as "the Antarctic Treaty, the measures in effect under that Treaty, its associated separate international instruments in force and the measures in effect under those instruments".

A.D. Hemmings (✉)
Perth, WA, Australia
e-mail: ahe30184@bigpond.net.au

A.D. Hemmings
Gateway Antarctica Centre for Antarctic Studies and Research, University of Canterbury, Christchurch, New Zealand

P.S. Goel et al. (eds.), *Science and Geopolitics of The White World*,
DOI 10.1007/978-3-319-57765-4_2

in turn issues round sealing,[3] marine harvesting,[4] mineral resources[5] and environmental protection.[6] Membership of the Antarctic Treaty has increased from the 12 original signatories of 1959 to 29 Consultative (i.e. decision-making) and 24 Non-Consultative Parties in March 2017. The Commission for the Conservation of Antarctic Marine Living Resources comprises 24 states plus the European Union as Members (decision-making), plus 11 Acceding States. Overall, 59 states are Parties to one or more instruments of the ATS, 31 of them decision-making Parties. The top-tier members of the ATS include the major states of both the Developed World and the Global South (Brazil, China, India, and South Africa) and, inter alia, the Five Permanent Members of the UN Security Council. Antarctica is no longer an item of high profile contention at the UN (Hemmings 2014).Whilst international environmental NGOs are calling for Antarctic marine protected areas, this is not activism at the levels of the earlier "World Park" advocacy; the worst excesses of IUU fishing have been reined in (Österblom et al. 2015); and Antarctica remains the only place on the planet generally free of crime and violence.

So, on the face of it, Antarctica is in good shape, its governance well embedded, robust, and adequate to the needs. This is indeed invariably the formal view from within the system itself. The 2009 *Washington Ministerial Declaration on the Fiftieth Anniversary of the Antarctic Treaty* unsurprisingly adopts a positive tone, reaffirms the historic verities of the ATS and betrays not an iota of concern about any deficiencies or strategic challenges before that system.[7] In a world of endless troubles locally, regionally and globally, and more immediately pressing concerns for governments, the Antarctic plainly appears a refreshingly benign geopolitical and institutional arena. Where else might one point to a system of international governance still ticking over so nicely 58 years after it was set in motion?

In this paper I want to challenge this happy picture of a benign and unproblematic contemporary Antarctic scene. In doing so the paper neither claims nor requires that Antarctica presents in a state of chaos or mayhem comparable to other areas of legitimate and necessary concern on our planet. It does not intend to suggest that the problems it identifies in our current Antarctic governance arrangements are of the same order as those in other places and around other issues. Its purpose is to seek to identify problems in an early stage of development, at a point where they are (if we are persuaded) far more amenable to resolution than they likely will be if left until they cross critical thresholds. The paper takes it as

---

[3][CCAS] Convention on the Conservation of Antarctic Seals. Adopted London, 1 June 1972, entered into force 11 March 1978. 1080 UNTS 175.

[4][CCAMLR] Convention on the Conservation of Antarctic Marine Living Resources. Adopted Canberra 20 May 1980, entered into force 7 April 1982. 1329 UNTS 47.

[5][CRAMRA] Convention on the Regulation of Antarctic Mineral Resource Activities. Adopted Wellington 2 June 1988. 30 ILM 1455. Has not entered into force, being superseded by the Protocol on Environmental Protection to the Antarctic Treaty.

[6][Madrid Protocol] Protocol on Environmental Protection to the Antarctic Treaty. Adopted Madrid 4 October 1991, entered into force 14 January 1998. 30 ILM 1461.

[7]ATS (2009, 161–162).

axiomatic that Antarctic governance should continue to manage emergent issues so that they do not become contentious,[8] and not await the arrival of crisis before addressing issues.[9] Finally, the approach adopted in this paper reflects the "new" and "deliberative" Antarctic exceptionalism that I believe is essential if we are to successfully preserve Antarctica and the common interests of humankind there into the future.[10]

2 'Hollowing Out' as a Concept

The concept of 'hollowing out' has been widely applied over the past decades in domestic political contexts—in relation to substantive changes in administration and public services (Rhodes 1994), and more recently in relation to the state of democracy in modern states (Mair 2013; Anheier 2015). The usage, naturally, varies in the detail, but the commonality is an attempt to capture a sense of the loss of the substantive capacity within a formally unchanging architecture and/or rhetoric. The authors deploying the 'hollowing out' rubric point to the gap between the formal and the actual state of affairs.

I have previously argued, in the context of the 50th anniversary of the Antarctic Treaty, that 'hollowing out' captures nicely at least some of the difficulties and challenges now faced by the ATS as a regional governance regime (Hemmings 2010) and the present paper offers a further development of this thesis. However, in important ways as discussed below the ATS has always been a hollow system, and it is therefore important to discriminate between foundational and structural hollowness and the more recent hollowing argued here. And by 'recent', I am largely talking about the period since the adoption of the Madrid Protocol in 1991.

The peculiarities of the Antarctic situation—variously attaching to the place as a geographical location; as a consequence of historically contingent issues around the onset and nature of Antarctic activity; the states that were in a position to conduct that activity; the assertions of claims by some states; and global framing such as the Cold War—had consequences. It meant that the formal edifice of the international institution elaborated to provide international governance of this contested space would necessarily be a structure both lightly built and somewhat isolated—sui generis in important ways. The sensitivities around territorial sovereignty, which were complicated by the alignment of the claimants with the Western bloc led by the United States, made this almost inevitable. Antarctica was only internationalized up to a point. A very high level of national autonomy was required, not only by

[8]The approach historically taken in relation to sealing, marine harvesting and minerals resource activities in the ATS.

[9]The first recital of the Preamble to the Antarctic Treaty refers to the "interest of all mankind" that Antarctica "not become the scene or object of international discord".

[10]On the challenges to the historic Antarctic exceptionalism, see Hemmings (2007, 2009). The argument for a new deliberative exceptionalism is presented in Hemmings (2009, 71).

the claimants but by the rival superpowers of the United States and Soviet Union, in order for the Antarctic Treaty (and subsequently added instruments) to be acceptable. This was achieved by first establishing consensus as the basis for decision-making—thereby enabling a de facto 'veto' by any Consultative Party—and thereafter through a clear but informal understanding that certain sorts of issues (most notably territorial and jurisdictional issues) would not be further litigated-within the Antarctic fora.

In a place so remote from metropolitan territories, without resident populations in the conventional sense being present in Antarctica, with its highest forum an annual meeting of officials as distinct from a meeting of Ministers who could address the essentially political issues that arose (and, as the ATS evolved, issues transcending the separate ATS instruments), there was from inception a 'democratic deficit' of a quite profound sort.

The original governance structure also did not cover the entire field of Antarctic activities. To take just one facet of this, consider the situation in relation to what has become a dominant theme within the ATS—environmental protection. The Antarctic Treaty left high seas freedoms in the region intact, and whaling to regulation by a global instrument, the International Convention for the Regulation of Whaling,[11] and its International Whaling Commission (IWC). Subsequently the ATS has adopted a practice of complete subsidiarity to the IWC,[12] beyond that necessarily required under the letter of the various ATS instruments (Rothwell et al. 2009). Furthermore, as the ATS developed, its modus operandi was to assign responsibility for managing particular issues to successive instruments, without prejudice to existing instruments. Whilst there were sound practical and geopolitical rationales for this approach, in conjunction with the extra-ATS placement of responsibility for whales, it has posed some interesting issues around how the protection of the Antarctic environment as a whole (central to the purposes of both CCAMLR and the Madrid Protocol), and the particular responsibilities of Antarctic Treaty Consultative Parties in relation to this protection, can be achieved in practice, with the resulting fractionated institutional responsibility (Hemmings 2013).

Resolving the deep structural hollowness of the ATS would require a willingness to conduct a major reassessment of the purposes and modalities of national Antarctic engagement and international governance of the region. This would necessarily require reconsideration of positions around the question of territorial sovereignty in Antarctica. There is no current prospect of such reconsideration, the original structural hollowness of the ATS will therefore not be addressed any time soon, and this paper does not hereafter consider this further.

---

[11]International Convention for the Regulation of Whaling. Adopted Washington 2 December 1946, entered into force 10 November 1948. 161 UNTS 74.

[12]See e.g. the consideration of environmental impact assessment duties in Antarctica in Hemmings et al. (2007).

3 Contemporary Hollowing of the Antarctic Treaty System

Institutionally, the ATS developed through decadal accretion of new topic-specific instruments addressing issues (essentially resource issues) as they arose: sealing through CCAS in 1972, fishing through CCAMLR in 1980, mining through CRAMRA in 1988, and generic environmental protection as well as the closing of the mining option, once CRAMRA was abandoned, by the Madrid Protocol in 1991. The approach of the ATS was to assume an exclusive responsibility for each emergent issue in the area 60 degrees South, or south of the Antarctic Convergence in the case of marine harvesting under CCAMLR. With the exception of the Antarctic manifestation of global whaling, noted previously, the ATS sought to own and regulate any substantive in-area activities. This approach reflected the historic Antarctic exceptionalism with which both the system and the stances of participating states brought to the region (Hemmings 2007, 2009).

A conjunction of events including the ending of the Cold War, improved operational capacity through technological advances and their wider dissemination, the rise of globalism and transformation of some states into quasi 'market states' (Bobbitt 2002), the dramatic expansion of human activities not just in Antarctica but surrounding it, and the growth in the number of global instruments that also apply in Antarctica, has now undermined this *modus operandum*. Emergent Antarctic issues are now, seemingly, construed by Consultative Parties as either inherently less problematical (and thus do not require new dedicated instruments for their regulation), or merely regional manifestations of a global activity (and thus best left to either global regulation, or the alleged sufficiency of market forces and/or self regulation).

The pattern of past ATS development may also have had influence. Because each successive additional instrument in the ATS was without prejudice to pre-existing instruments, as the system built the ATS acquired internal inconsistencies and complexity. The difficulties naturally increase with the number of instruments, and possibly the growth-by-accretion model has anyway reached its natural limits. For a community anxious about the risks of the alternative—a general system re-jigging, wherein a Pandora's Box of uncertainties might be opened up—this may now also be a disincentive to further ATS development. But the substantive change is in the global context, where the historic Antarctic regional approach is now orphaned. The ATS as a body has yet to formally determine whether it is content to allow this to continue, or whether (and if so, how) it might establish a new basis for the regional management of Antarctica.

But, the ATS has adopted no new substantive instrument since the Madrid Protocol in 1991. The now over a quarter century without substantive addition to the ATS is its longest period without any instrument being under development. One might ask whether, quite aside from the specific 'gaps' in regime coverage this has allowed (below), the absence of a process of negotiation has also denied the ATS a critical structural 'glue'. The process of negotiating something necessarily involves states interacting in a dynamic context. It brings in new players (new states that are interested in this regulatory instrument; new communities in established Antarctic states who see themselves as stakeholders) and it generally alters and reinvigorates bureaucratic engagement within existing ATS states. On this last point, put cruelly, significant new diplomatic negotiation often clears out the 'dead-wood' and brings more senior and capable people into national delegations.

The hollowing of Antarctic governance is reflected in several 'paths-not-taken' since the adoption of the Madrid Protocol. Three will be examined here. These are (a) in relation to *regulatory gaps*—where new (or more intense) human activities, organized around actual or incipient industries, have not been regulated by the ATS in the traditional manner, (b) in relation to high profile—one might say 'strategic' policy—issues concerning territorial sovereignty, tied up in the question of the *Antarctic continental shelf* and (c) in relation to the operation of one of two main diplomatic fora of the ATS i.e., the annual Antarctic Treaty Consultative Meeting.

3.1 Regulatory Gaps

In terms of 'gaps', two obvious candidates present. Antarctic tourism and bioprospecting are the two industries which have largely developed in the period since the adoption of the Madrid Protocol.[13] Whilst subject to the generic obligations of the Antarctic Treaty and (particularly) the Madrid Protocol (and in relation to bioprospecting also CCAMLR), neither has been subject to the sort of response that earlier emergent or potential activities/industries (sealing, fishing, mining) or even long established whaling (outside the ATS) elicited. We have not seen the development of a 'Convention for the Regulation of Antarctic Tourism' or a 'Convention for the Regulation of Antarctic Biological Prospecting'. Neither have we seen substantive regulation of these activities through second-order instruments (legally binding Measures under the Antarctic Treaty/Madrid Protocol; Conservation Measures under CCAMLR). Thus, although Measures in relation to tourism were adopted in 2004 and 2009[14] neither has yet entered into force. In a recent major review of the Antarctic Treaty Consultative Meeting's (ATCM) consideration of Antarctic tourism, India (2015) has identified the substantial discussion of tourism that has occurred between the Consultative Parties, noting that some

> Issue-areas related to the regulation of Antarctic tourism have stayed on the ATCM agenda and in some cases repeatedly discussed and highlighted over several decades… [and] that it is not so much a question of whether but when the ATCPs would turn to a more focused discussion of how best to formalize, institutionalize and operationalize the insights, resolutions, recommendations (including those mentioned in Annexes A, B and C to this Information Paper and measures that have accumulated over the decades at various ATCMs.

[13]On tourism see; Liggett et al. (2011) on bioprospecting see Leary (2014).

[14]Measure 4 (2004) Tourism and Non-Governmental Activities; Measure 15 (2009) Landing of Persons from Passenger Vessels.

The problem is not just confined to tourism. The last time an ATCM adopted any Measures on anything apart from those associated with Protected Area management plans was in 2009, when in addition to Measure 15 on tourism, Measure 16 on the amendment of Annex II to the Protocol was adopted (the latter has not entered into force either). To put this in context, the seven ATCMs since Baltimore (i.e. 2010–2016) have adopted 103 Measures.In short, over recent years, the ATCM has ceased to adopt legally binding Measures on anything apart from Protected Area Management Plans.[15] Whilst hortatory Resolutions undoubtedly have their place, the absence of an evolving body of legally binding obligations leaves the Antarctic Treaty and Madrid Protocol with increasingly dated and limited tools across its broader field of competence. Under CCAMLR, the analogous Conservation Measures have continued to be annually adopted across a range of operational management tasks. The problem-area in relation to CCAMLR in recent years has been particularly around the failure to reach consensus on the designation of marine protected areas. Whilst this circumstance poses significant challenges, it is not so clearly evidence of 'hollowing' per se.[16] Whatever the arguments for or against 'Convention' equivalent instruments for these later commercial activities, their absence means that increasing parts of current Antarctic activity is not substantively regulated by the ATS. There are some informal mechanisms, but most of these devolve to relationships between industry and particular states.

3.2 *Antarctic Continental Shelf*

Coastal states are entitled under Article 76 of the UN Convention on the Law of the Sea (UNCLOS)[17] to certain rights over the continental shelf beyond 200 nautical miles if they can demonstrate its extent through data submitted to the Commission on the Limits of the Continental Shelf (CLCS). The seven territorial claimants in Antarctica see themselves as coastal states sensu UNCLOS, and have variously sought to reserve their rights as such in relation to the continental shelf appurtenant to their Antarctic claims. The details of the particular decisions taken by claimant states in relation to the extended continental shelf (ECS) in the Antarctic Treaty area have been well worked in the literature.[18] The issue here is that over the past seventeen years[19] despite the continental shelf having been a major Antarctic

[15]Discussed in the annual review of The Antarctic Treaty System in New Zealand Yearbook of International Affairs since 2011. See, e.g. Hemmings (2017a).

[16]Although the underlying differences around the interpretation of purposes of the convention may be. See Cordonnery et al. (2015).

[17]United Nations Convention on the Law of the Sea. Adopted Montego Bay 10 December 1982, entered into force 16 November 1994. 21 ILM 1261.

[18]Succinct coverage provided in Saul and Stephens (2015), lxiii–lxvi.

[19]The period may reasonably be said to begin in 1999 with an Australian Media Release: Robert Hill, Minister for the Environment and Heritage and Alexander Downer, Minister for Foreign

geopolitical issue, raising legal questions relating to territorial positions (Kaye 2001) and stimulating six non-claimant ATCPs to lodge notes with the CLCS reiterating non-recognition of territorial claims and the provisions of Article IV of the Antarctic Treaty,[20] absolutely no consideration of it has occurred in the fora of the ATS.[21] There is no reference to the issue in the Final Reports of any ATCM. To an objective observer, the absence of formal consideration of such a significant issue in the very system established to manage Antarctic interests is surprising. It is ironic that one of the core issues behind the negotiation of the Antarctic Treaty, and seemingly contained by it through the celebrated Article IV, is now so sensitive that it cannot any longer be placed on an ATCM agenda and discussed. The related question of jurisdiction in Antarctica has also proven too sensitive to consider.

3.3 Antarctic Treaty Consultative Meeting

Article IX (1) of the Antarctic Treaty itemizes "measures in furtherance of the principles and objectives of the Treaty" which the Representatives of Parties may wish to consider at subsequent meetings (the meetings that we refer to as ATCMs). Amongst these are "(e) questions relating to the exercise of jurisdiction in Antarctica". Whereas the other measures itemized have, to varying extents, indeed provided the basis for subsequent ATCM discussion and specific outcomes, the question of jurisdiction was left alone, until the 1992 ATCM, when Uruguay tabled a Working Paper under the title: "Issues relating to the exercise of jurisdiction in Antarctica" (Uruguay 1992).

This paper was stimulated by a January 1992 incident involving a Uruguayan and the death of a Russian at the Russian station on King George Island. Summarizing the juridical complexities around this incident (which the Uruguayan paper itself did not explicate): we had a citizen from a non-claimant ATCP (Uruguay,) involved in the death of a citizen of a state (Russia) which rejects territorial claims but asserts that it has a basis to claims itself, at a Russian station, in a part of the Antarctic subject to mutually exclusive territorial claims by three states (Argentina, Chile, United Kingdom)—none of which claims Uruguay has publicly recognised.

Although an agenda item was established for the ATCM: "Item 18. Question Related to the Exercise of Jurisdiction in Antarctica", it was not substantively

(Footnote 19 continued)

Affairs. "Move to Claim Extended Antarctic Continental Shelf"—Joint Media Release, 2 December 1999.

[20]Germany, India, Japan, Netherlands, Russian Federation, United States. Data from http://www.un.org/depts/los/clcs_new/commission_submissions.htm.

[21]See the statement by a senior Swedish diplomat that although the claimants, Russia and the US had held consultations, "no formal or informal consultations with the rest of the Parties were held" (Jacobsson 2007).

discussed and the ATCM decided that it would be considered at the next ATCM, which was not held until 1994. At that meeting, held in Seoul, in the space of two days, two revisions of the paper were issued, excising reference to the death incident and softening the analysis. Following a brief discussion the ATCM Final Report (Republic of Korea 1995) noted:

> (122) A working paper (XVIII ATCM/WP 32) on this item was tabled and introduced by Uruguay. The Meeting recognized the importance of this question, the solution of which was left deliberately open in Article IX (1) of the Antarctic Treaty. But it was also understood that the question raises some delicate and sensitive problems which need more, and careful, deliberations.
>
> (123) The Meeting therefore agreed to leave the item out of the Agenda of the XIXth ATCM and put it again on the Agenda of the XXth ATCM in order to give all Parties sufficient time to elaborate ways and means how to approach the question again in order to find an agreeable solution.

Accordingly, the issue was put onto the agenda of the XX ATCM held in The Netherlands in 1996. There was no substantive discussion and the Final Report included two short paragraphs (Netherlands 1996), the first reiterating the sentiments of paragraph 123 of 1994, and the second reading:

> (74) The Meeting agreed that the Delegations had not yet had sufficient time to duly consider the issue, and decided to omit the item from the Agenda of the following Consultative Meetings until a request was made by a Consultative Party to re-include it.

So, a paper raising issues around jurisdiction (seemingly recognized at the adoption of the Antarctic Treaty in 1959 as an issue that needed attention), triggered by a serious incident in the Antarctic, was reluctantly passed through three ATCMs over a period of four years before being put out to pasture. The item has never been revisited, and the general problem of jurisdiction in such circumstances remains unresolved, beyond the hortatory 2012 injunction that "the Parties cooperate to institute discussion on issues related to the exercise of jurisdiction in the Antarctic Treaty area",[22] which rather begs the question. One might argue that, notwithstanding the text of Article IX, this is an example of the foundational structural hollowness of the Antarctic Treaty. But this seems a weak position. Great care surrounded the drafting of the Treaty, and it seems implausible that Parties then deliberately identified an issue for subsequent attention which they had no intention of considering. The alternative reading that the unwillingness to consider the matter in the mid 1990s is in some sense a product of a latter hollowing of the ATS seems, on the face of it, more probable.

The other facet of the ATCM that I want to address here is the decision to contract the length of the annual meeting from ten to eight working days. The decision to do this was made at the 33rd ATCM in 2010 (ATS 2010) although this is documented only in the Final Report at paragraph 537, and not codified as a Decision (the class of decision-making ordinarily used in relations to matters of an

[22]Resolution 2 (2012).

administrative nature),[23] and scheduled to commence at the next-but-one ATCM, i.e. the 35th ATCM in 2012 in Hobart. Plainly, there had been prior discussion between at least some Consultative Parties on this matter. Formally, the rationale for this contraction in the duration of the meeting was "efficiency" in the context of the development of a "Multi-Year Strategic Work Plan" for the ATCM (ATS 2010 paragraphs 519–527), the lead on which was Norway, which "introduced a proposal". Norway had not tabled any paper containing such a proposal—and nor had any other State or participant (Observer or Expert). For a substantive proposal, one might ordinarily have expected a Working Paper. The decision to hold an eight day meeting in 2012 was formally without prejudice to subsequent meetings: "as far as subsequent ATCMs are concerned, the Meeting further agreed that the appropriate length of ATCMs would be kept under review" (ibid, paragraph 538). But, at the following 34th ATCM, Norway (2011) tabled a Working Paper specifically proposing the shortening of ATCMs, and at the first of the shorter ATCMs (the 35th in Hobart) Belgium announced the dates for the 36th ATCM showing that it planned to continue the eight day meeting format (ATS 2012, paragraph 308).The following year, the next ATCM (host, Brazil) also confirmed an eight day meeting (ATS 2013). This pattern has continued with the 2015 and 2016 ATCMs.

Plainly, the eight day ATCM is established as the new norm. Aside the question whether a degree of stealth was involved at the inception of this process—and of how many Consultative Parties were privy to the move—the substantive question I want to pose is, does this materially strengthen or weaken the ATCM's capacity to conduct its business and deliver on the obligations of the two international legal instruments (the Antarctic Treaty and the Madrid Protocol) that it purports to manage?

All the indications are that the environmental management obligations of the Madrid Protocol (to say nothing of the wider obligations that arise under the Antarctic Treaty) are expanding, rather than contracting over the years. This is the unremarkable product of increasing human activity in Antarctica, more participant States within the ATS, a global trend of more sophisticated environmental management, and the learning curve and capacities in relation to Antarctic activities in particular. Whatever the 'efficiencies' achieved through the entirely reasonable process of implementing work-plans for the ATCM and its advisory Committee for Environmental Protection, it seems inherently unlikely that cutting the yearly meeting by 20% is going to be without effect on its capacity to not only get through the agenda, but give the issues the substantive consideration they may often require. Might the disappearance of the legally binding Measure from the output of recent ATCMs (except in the case of Protected Area management plans) be coupled with this? Of course these may be two separate symptoms of a more general shift of focus. In either case, this is a facet of the hollowing of Antarctic governance that we have under review here. Which takes us to the key question: What has caused this post-Madrid Protocol hollowing of Antarctic governance?

[23]Decision 1 (1995).

4 Drivers of the Hollowing of Antarctic Governance

The Madrid Protocol was adopted after not only a particularly difficult period for the ATS (the decade long negotiation and then abandonment of the minerals convention CRAMRA; the associated critique of the ATS by global environmental NGOs and large parts of some members' publics; the challenge to ATS hegemony constituted by the 'Question of Antarctica' in the UN General Assembly) but at a hinge point in global geopolitics, when the Soviet system collapsed and new (if uncertain) world orders were possible. Even in Antarctic fora the transformation was palpable. At my first ATCM (Paris in 1989), Consultative Parties still aligned themselves by reference to their superpower sponsor, or in the case of non-aligned states by conspicuous detachment. I recall my then head of delegation, renowned New Zealand diplomat Chris Beeby, when East Germany had supported New Zealand's then pro-CRAMRA stance, turning to me and wryly remarking that "things aren't quite going as we hoped!" By the time we got to Viña del Mar in late 1990 and the onset of the negotiations for what became the Madrid Protocol, the Soviet Union, whilst still formally an entity was a busted flush, and by the 1992 Seventeenth ATCM had been superseded by the Russian Federation. Everything had changed in Antarctic terms.

My first proposition is that when we look to explanations for the various Antarctic transformations—including the argued hollowing of its governance—we recognize the criticality of the first few years of the 1990s. Having just got through a traumatic (in relative terms of course) decade, the ATS was still intact, it had a new shiny instrument to both bring into force and learn how to operationalize (quite different things), and it had some outliers still to negotiate (a Secretariat; and some sort of arrangements around liability for environmental damage). But this was all assumed to be 'bedding-in' stuff. Antarctica was not going to be a 'hot' topic any more. So very probably we were always in for a quiet decade in Antarctic affairs. The difficulty was, that (a) hitherto the ATS had always had the proverbial 'pot on the boil', with instruments adopted each decade, and had therefore never experienced a 'quiet' period before; and (b) it was precisely during this quiet decade that the pattern of Antarctic activity changed most significantly, with the new and burgeoning industry of tourism joining the decade-old Antarctic fishing industry as Antarctica's new commercial players. It was not, in my view, that the tourism industry was necessarily a problematical development, but that for historically contingent reasons its arrival elicited none of the preparatory regulatory work (bar the encouragement that particular states gave to the establishment of IAATO) that the arrival of previous industrial interests (including fishing) had.

Second, my sense is that after the nasty surprise some states had had with the collapse of CRAMRA, there was an inclination to steer well clear of any further attempts to regulate anything. No official wanted to propose to their government that another can of worms be opened! For some states it was simply this aversion to risk. For others I think that a strategic choice was made to leave 'regulation' to informal or market mechanisms. That way one avoided formal responsibility for particular codifications, there was nothing to be overturned by your opponents, and thus the thing (generally commercial activity) was likely to be able to proceed. Much less fuss and bother all round.

So, we faced a situation where the existing (and newly agreed) elements of the ATS were all in place, but where there was no enthusiasm for engaging in new and potentially contentious regime development. As we know, the Secretariat was eventually resolved, and a limited annex on liability (Annex VI to the Madrid Protocol) was adopted in 2005, although twelve years on its entry into force is still years away at best, and may never be achieved.

As I have argued elsewhere, it also appears that we began to come up against the limits of what we actually agreed on in relation to the Antarctic dispensation by the turn of the century. The incremental development of the ATS, and its convention of not relitigating past arrangements, meant that we had sometimes arrived at consensus around quite vague ideas of the actual obligations. As the cumulative weight of these obligations increased, and more so as we actually began to operationalize them, some of these obligations suddenly became rather inconvenient. Whilst states have been largely free to self-define Antarctic obligations, there is a limit to the viability of this in a necessarily collective enterprise. Various matters have thrown up the quite different stances Consultative Parties may take in relation to particular issues—whether it is environmental impact assessment (Hemmings and Kriwoken 2010), the designation of marine protected areas (Cordonnery et al. 2015) or environmental management overall (Hemmings 2013). Furthermore, we seem to be having some general difficulties around the vexed subject of what values we are aiming to secure in our Antarctic arrangements (Hemmings 2012). In this respect we face not only the inevitably differing positions of states, but one otherwise positive achievement of the ATS—its longevity. This has meant that we are using instruments now all decades old to address contemporary situations; and unsurprisingly the values of the past are sometimes not those of the present—or at least they are not the preferred values of some states.

The contemporary Antarctic situation combines both the traditional inter-state contestation of Antarctic territorial sovereignty evident in the 1950s and the new technology-enabled frictions of actual and reasonably foreseeable commercial activities. Whereas, the pre-Antarctic Treaty problem was that territorial sovereignty issues (given further edge by their situation alongside the global ideological bipolarity of the Cold War) occurred outside any institutional framework, the present problem occurs despite the operational continuation of that framework, the ATS. Further, not only is the possibility of resource/commercial activity which was all but impossible in the 1950s now a reality, but formal sanction for the activities is present (or asserted to be) in both ATS and non-ATS global instruments. Further, under the influence of globalism and neoliberalism, it has proven increasingly difficult to agree any substantive Antarctic-specific governance system for local manifestations of emergent global industries (which include both tourism and bioprospecting).

Now that Antarctica is seen to possess actually realizable commercial benefits (including, notwithstanding the current minerals resource activity prohibition, under

Article 7 of the Madrid Protocol, the ongoing interest in hydrocarbons and, perhaps further out, other mineral resources), it is even harder for claimants to let go of their supposed territorial rights than it was in the past, and all potential beneficiary states are disinclined to adopt obligations that might come back to haunt them later. This is particularly so if global regimes promise to grant valuable rights to them precisely because they are territorial/coastal states (the UNCLOS Article 76 dilemma). Absent the Cold-War glue and the general technical restraints of a difficult Antarctic environment, in an international milieu of unrestrained commercial competition, the Antarctic regime is extremely vulnerable. It cannot grow; it cannot easily update even those capacities that it has; it faces competition from global regimes (or global norms of market liberalization outside traditional institutional regulation) for jurisdiction over particular fields; and its original problem (territorial sovereignty) is reinvigorated, and one route whereby the broader global phenomenon of nationalism may gain traction in relation to Antarctica too (Hemmings et al. 2015).

In consequence, the ATS has transformed from a regional regime wherein substantive policy responses were not only adopted, but also operationally managed, to one which risks becoming merely a limited regional coordinating mechanism, with substantive responses either impossible at all, or left to other fora. What we have seen is a disabling of the substantive core of the ATS, whilst leaving its edifice intact. It has not collapsed, but it has been hollowed out.

5 The Future

There are no easy remedies for this hollowing. The arrangements for Antarctic governance are those that sovereign states are prepared to countenance, and counsel from the sidelines has little chance of influence unless those states come to see their present approaches to Antarctica as problematical in terms that matter to them. But it would, surely, be desirable to see some more substantive discussions of core Antarctic governance issues and options than has been evident within the ATS since (probably) the original Washington conference before the adoption of the Antarctic Treaty. A hollowed Antarctic governance may be a present convenience in the estimation of some participants (although one might wonder how many Antarctic active states actually conduct any sort of high policy discussion around Antarctic governance—in which case, is what we see actually deliberative?) but it risks the real control of Antarctic futures migrating elsewhere, and the declaratory commitments to the ATS suggest most Consultative Parties would not wish to see this.

My own view is that we need to find a new basis for confidence in a regional governance arrangement in the Antarctic, and that requires new thinking as well as a confidence that some historic Antarctic values are worth defending (Hemmings 2014). There is an obvious double project open to the ATS: to frame its principles and values (and hence the underpinning of its justification) in a way that allows these to be 'owned' beyond the community presently found within the ATS, and to internalize global principles and values not hitherto reflected within the ATS so that the Antarctic finally sits more comfortably in the global context (Hemmings 2017b). If we can reinvigorate and expand the Antarctic Treaty System, so that it can continue to provide the institutional architecture to do this, I think this is our best option. But if it cannot or will not evolve, there is nothing about the Antarctic situation that will prevent the real power to determine Antarctic futures going elsewhere.

References

Anheier HK (2015) Conclusion: how to rule the void? policy responses to a 'Hollowing Out' of democracy. Glob Policy 6:127–129

ATS (2009) Final report of the thirty-second Antarctic treaty consultative meeting, Baltimore, United States, Secretariat of the Antarctic Treaty, Buenos Aires, 6–17 Apr 2009

ATS (2010) Final report of the thirty-third Antarctic treaty consultative meeting, Punta del Este, Uruguay, Secretariat of the Antarctic Treaty, Buenos Aires, 3–14 May 2010

ATS (2012) Final report of the thirty-fifth Antarctic treaty consultative meeting, Hobart, Australia, Secretariat of the Antarctic Treaty, Buenos Aires, 11–20 June 2012

ATS (2013) Final report of the thirty-sixth Antarctic treaty consultative meeting, Brussels, Belgium Secretariat of the Antarctic Treaty, Buenos Aires, 20–29 May 2013

Bobbitt P (2002) The Shield of Achilles: War, Peace and the Course of History. Allen Lane, London

Cordonnery L, Hemmings AD, Kriwoken LK (2015) Nexus and imbroglio: the Madrid protocol and designating antarctic marine protected areas in the Southern Ocean. Int J Mar Coast 30 (4):727–764

Hemmings A. (2007) Globalisation's cold genius and the ending of Antarctic isolation In: Kriwoken LK, Jabour J, Hemmings AD (eds) Looking South: Australia's Antarctic Agenda. Federation Press, Sydney, pp 176–190

Hemmings AD (2009) From the new geopolitics of resources to nanotechnology: emerging challenges of globalism in Antarctica. Yearb Polar Law 1:55–72

Hemmings AD (2010) After the party: the hollowing of the Antarctic treaty system and the governance of Antarctica. Paper presented at the symposium on Antarctic politics, New Zealand: University of Canterbury, Christchurch, 8–9 July. Available at https://www.researchgate.net/publication/281029050_After_the_Party_The_Hollowing_of_the_Antarctic_Treaty_System_and_the_Governance_of_Antarctica

Hemmings AD (2012) Considerable values in Antarctica. Polar J 2(1):139–156

Hemmings AD (2013) Environmental management' as diplomatic method: the advancement of strategic national interest in Antarctica. In: Liggett D, Hemmings AD (eds) Exploring Antarctic Values. Gateway Antarctica, Christchurch, Special Publication Series 1301, pp 70–89

Hemmings AD (2014) Re-justifying the Antarctic treaty system for the 21st century: rights, expectations and global equity. In: Powell RC, Dodds K (eds) Polar geopolitics? Knowledges, resources and legal regimes. Edward Elgar, Cheltenham, pp 55–73

Hemmings AD (2017a) The Antarctic treaty system (The Year in Review 2015). N Z Yearb Int Law 13:272–281

Hemmings AD (2017b) Antarctic politics in a transforming global geopolitics. In: Dodds K, Hemmings AD, Roberts P (eds) Handbook on the Politics of Antarctica, Cheltenham: Edward Elgar, pp 507–539

Hemmings AD, Kriwoken LK (2010) High level Antarctic EIA under the Madrid protocol: state practice and the effectiveness of the comprehensive environmental evaluation process. Int Environ Agreem: Polit Law Econ 10(3):187–208

Hemmings AD, Chaturvedi S, Leane E, Liggett D, Salazar JF (2015) Nationalism in today's Antarctic. Yearb Polar Law 7:531–555

Hemmings AD, Scott KN, Rogan-Finnemore M (2007) Broadening the Duty in relation to environmental impact assessment across the legal instruments applying in Antarctica. Paper presented at the 15th annual conference of the Australian and New Zealand Society of International Law. Restoring the rule of law in international affairs, Australian National University, Canberra, Australia. 28–30 June. Available at http://www.worldlii.org/int/journals/IHLRes/2007/11.html#fn1

India (2015) Towards a comprehensive, proactive and effective Antarctic tourism policy: turning recommendations into action. Information Paper 104 rev. 1. Thirty-eighth Antarctic Treaty Consultative Meeting, Sofia, Bulgaria, 1–10 June

Jacobsson M (2007) The Antarctic treaty system: legal and environmental issues—future challenges for the Antarctic treaty system. In: Triggs G, Riddell A (eds) Antarctica: Legal and Environmental Challenges for the Future. British Institute of International and Comparative Law, London, pp 1–16

Kaye SB (2001) The outer continental shelf in the Antarctic. In: Oude AG, Elferinkand AG, Rothwell DR (eds) The Law of the Sea and Polar Maritime Delimitation and Jurisdiction. Martinus Nijhoff, The Hague

Leary D (2014) From hydrocarbons to psychrophile: the 'scramble' for Antarctic and Arctic resources. In: Stephens T, Vander Zwaag DL (eds) Polar Oceans Governance in an Era of Environmental Change. Edward Elgar, Cheltenham, pp 125–145

Liggett D, McIntosh A, Thompson A, Gilbert N, Storey B (2011) From frozen continent to tourism hotspot? five decades of Antarctic tourism development and management, and a glimpse into the future. Tour Manag 32:357–366

Mair P (2013) Ruling the Void: The Hollowing of Western Democracy. Verso, London

Netherlands (1996) Final report of the twentieth Antarctic treaty consultative meeting. Ministry of Foreign Affairs, The Hague

Norway (2011) Proposal for shortening the Antarctic treaty consultative meetings. Working paper 60, Thirty-fourth Antarctic treaty consultative meeting, Buenos Aires, 20 June–1 July

Österblom H, Bodin Ö, Sumalia UR, Press AJ (2015) Reducing Southern Ocean: a global effort. Solutions 4(5):72–79. Available at http://www.thesolutionsjournal.org/node/237225

Republic of Korea (1995) Final report of the nineteenth Antarctic treaty consultative meeting, Ministry of Foreign Affairs, Seoul

Rhodes RAW (1994) The hollowing out of the state: the changing nature of the public service in Britain. Polit Q 65(2):138–151

Rothwell DR, (Chair), Stephens T (Rapporteur), Hemmings AD, Kaye S, Mossop J, Triggs G (2009) Japan's 'scientific' whaling program and the Antarctic treaty system—independent panel of legal and policy experts: report of the Canberra Panel. International Fund for Animal Welfare, Sydney, 12 January

Saul B, Stephens T (eds) (2015) Antarctica in International Law. Hart, Oxford

Uruguay (1992). Issues relating to the exercise of jurisdiction in Antarctica. Working paper 17 tabled at XVII Antarctic treaty consultative meeting, Venice, 11–20 Nov

State Practices on Antarctica and International Law: Attempt at Identification of India's Interests, Needs and Future Position

Bimal N. Patel

Abstract International Law governing the Antarctica Treaty Regime is heavily influenced by the evolving state practice than the true interpretation of the provisions of the Antarctica Treaty of 1959. The Treaty reflecting the power structures, circumstances and global affairs has undergone sea change since its entry into force. Like other international treaty regimes, non-influential participants have started taking keen strategic, environmental, economic and research interests—the trends which are nothing but consolidating. India and China then developing countries, now the important powers in the world, are ensuring that their needs, interests and concerns are reflected into any legal and policy norms and decisions. India, although a latecomer in the Antarctic diplomacy, is steadily influencing and contributing to shape the evolving norms and practices, in particular. Indian position is more likely to be shaped in future by the positions of her Asian neighbours, but at the same time, India will also carve out its own indigenous profile that meets her future political, economic, environmental and research interests. This paper attempts to examine and understand the state practices of major Antarctic powers in the areas of bio-prospecting, territorial claims, IUU fishing (Illegal, Unregulated and Unreported), fishing environment and climate change, oil and mining, mineral exploration, governance, United Nations Convention on Law of the Sea (UNCLOS) and continental shelf, tourism, liability and dispute settlement. It puts the Indian state practice in this overall framework and sees how India, with its emerged power status, is and will influence the evolving norms.

Keywords Antarctic treaty · State practice · Bio-prospecting · Liability regime · Strategic interests

---

B.N. Patel (✉)
Gujarat National Law University, Gandhinagar, India
e-mail: vc@gnlu.ac.in

P.S. Goel et al. (eds.), *Science and Geopolitics of The White World*,
DOI 10.1007/978-3-319-57765-4_3

1 Introduction

The Antarctic Treaty (AT) consists of 2364 words in preamble and 14 articles, and as such is one of the shortest international instruments put in the context of a regime which it governs. No other contemporary international law regime is as complicated and yet has generated very few real conflicts as the Antarctic Treaty System (ATS). All possible challenges of running a nation, a region and international space are present. The issues and challenges include sovereignty (Scott 2011), resource protection (Davies 2013), governance, regime compliance and the problem of how to manage the activities of commercial operators based in countries which are non-signatories. Even there are talks of a new multilateral tourism convention on Antarctic level (Bates 2011).

It is important to understand the operation of regime from an overall perspective of various member States as their overall interests and approach determine the governance of the ATS. For example, South Korea has strong geostrategic, scientific, environmental, economic and political interests. It approaches the ATS heavily from political and economic interests and improving the nation's international status. USA, while ratifying the Treaty, considered that the ATS was required to protect its immediate and long-term interests and also the ATS was seen as a good precedent in the field of disarmament, prohibition of nuclear explosions and law of space. New Zealand in addition to strategic and political interests has well-laid down policies on tourism, fishing, and logistics for national and international expeditions. China is emerging as a major power in Antarctica (Brady 2013). Not only China has developed ice-capable planes, enabling researchers from China to fly into Antarctic, rather than go by boat but its extensive program of international collaboration with Canada, Chile, France, Germany, Norway, Japan, New Zealand, South Korea, Romania, Russia, UK and US, show how Chinese bilateral and multilateral collaboration will reap benefits to the nation but as well as give her crucial insights into the weaknesses and strengths of internal regimes of partner nations. Apparently, India is missing in international collaboration network of China. China has huge interests in Antarctic resources and any possibilities for their exploitation. For China, minerals, meteorites, and the intellectual property of bio prospecting, locations for scientific bases, marine living resources and access to the continent for Antarctic tourism, all are areas of significant long-term interests. In fact, the Chinese rise has attracted international apprehensions too but China keeps expanding its interests in Antarctica even at the risk of its image. A question arises, is India willing to risk its image in Antarctica like China? Against this emerging state-interest centric operation of the regime, let us examine developments in important issues and see where India stands and what position India ought to take.

2 Bio-prospecting

Bio-prospecting is currently unregulated and therefore becomes a case study to re-examine the fundamental question as to whether the Antarctica is global commons and whether its resources should be considered the common heritage of mankind (CHM). While analyzing the CHM and bio-prospecting, the concepts of equity, international public good and early and effective intervention from like-minded nations are imperative (Joyner 2012). Bio-prospecting is not simply a scientific and economic issue. It has international legal and political importance. It can remind us how the informal assertion on the part of the claimant states in relation to the extended continental shelf in the Antarctic treaty area (Weber 2008) are discussed and realised on the margins of the ATS rather than through the mainstream channels of Antarctic diplomacy such as the Antarctic Treaty Consultative Meeting (ATCM).

Nearly 2000 research organizations and companies, many of which are America based, are engaged in preliminary research on living resources for commercial purposes. Several complications seem apparent should bio-prospecting become a pervasive activity. Do claimant states possess the legal authority to regular access of commercial bio-prospectors? Would this authority give claimant states the right of refuse to a prospective bio-prospector? Who retains the profits, if any, that are derived from bio-prospecting research—the companies, the claimant states, the Antarctic Treaty Consultative Parties (ATCPs) or the international community? Would claimant states enjoy that same right under an Antarctic bio-prospecting regime? Regarding the freedom of scientific research within the treaty area, should a distinction be made between basic scientific researches, applied scientific research and commercial use of resources within the context of benefit sharing? How a member State would be able to remain engaged in monitoring the escalation of bio-prospecting activities and how would they prevent and control potential conflict among the ATCPs?

As we all know, although ATCPs have utilized the World Heritage Convention to nominate and manage areas on sub-Antarctic islands, they have resisted the extension of this instrument to Antarctica. Debates on biological prospecting, the study of genetic materials and determination of commercially important genetic codes and the harvesting of an in situ organism for extraction of biochemical product, within the ATCM have been resisted, linking this matter to programmes and practices with the United Nations Environmental Program (UNEP). Thus, a long-term focused attention on part of India, as to how this regime is evolving is imperative.

3 Territorial Claims and Issues of Jurisdiction

Territorial claims and issue of sovereignty constitute important legal framework whoever analyses the ATS. Indian stand is clear that it does not recognise any state's right or claim to territorial sovereignty in the Antarctic area and consequently over the seabed and subsoil of the submarine areas adjacent to the continent of Antarctica. Prior to ATS, 7 states had a historical claim to Antarctic territory namely, Australia, France, New Zealand, Norway, UK, Argentina and Chile. 29 states are having consultative status and 23 are non-consultative Parties. These claims are based on discovery and effective occupation of the claimed area, and are legal according to each nation's laws. Three countries—the United Kingdom, Chile and Argentina—have overlapping claims in the Antarctic. Australia and New Zealand claim mainly rest on the geographical proximity, while Argentina and Chilean claim rest on geographical extension of their territory, whereas Australia, New Zealand, France, Norway and UK would like to play significant role in Antarctic exploration. Apparently, in addition to Australia itself, four countries recognise the Australian Antarctic Territory (AAT). Whether then other nationals visiting so-called AAT ignore the Australian laws relating to fishing, by catch mitigation and whale hunting? The AAT cover nearly 5.9 million km^2, which is nearly 42% of Antarctica and about 80% of the Australia itself.

As we all know the Antarctic Treaty has put aside the potential for conflict over sovereignty by providing that nothing that occurs while the Treaty is in force will enhance or diminish territorial claims. Member states cannot make any new claims while the Treaty is in force. Meanwhile, the countries active in Antarctica meet every year to discuss issues as diverse as scientific cooperation, measures to protect the environment and operational issues. They are committed to taking decisions by consensus, and have all made the commitment that Antarctica should not become the scene or object of international discord.

What is the state practice in this regard? USA does not recognize the validity of sovereign territory on the continent from which offshore jurisdictional claims might be asserted. The critical point for US national interests in preserving the treaty would come if a claimant government were to act to effect its offshore claims through the proclamation of executive decrees or adoption of national legislation to that end. In passing reference, it may be worth mentioning that the US station Amundsen-Scott is located at the point of intersection of the territorial claims that US does not recognize. As a consequence, the USA occupies the apex of Antarctica, a symbiotically critical location where all the claims converge on the continent. If we look at the Russian practice, then we observe that Russia has planted its flag at the point of convergence of all the world meridians. The former Soviet Union also established research stations in each of the claimant states territories during the international geophysical year in 1957–58. Furthermore, majority of all year-round Russian stations are located in the AAT causing concern to Australia that the Russian, Chinese and Indian stations in AAT would lessen Australia's influence. Russia would like to maintain the current form of Antarctic Treaty (AT); however, it faces a challenge as to how to meet the possibility of its future inability to negotiate a continuation of the AT and the consequent floor of claims to Antarctic territory, given especially declining funding of Antarctic programs in Russia in the last two decades. The Chinese and Indian state practice show that they have built new bases on the territory that Australia claims.

Claimant states, their interest protected by the provisions of article IV of the treaty, have been able to balance the interests and their treaty commitments through what has been termed bifocalism, which refers to the productively ambiguous formation of the question of sovereignty allowing claimant states to act on the basis of territoriality while non-claimants act under article IV on the basis of their status as consultative parties.

The issue of exercising jurisdiction is very complicated. In 1959, the negotiators apparently left it to the state practice to develop tools to deal with this issue. Accordingly, article VII (5) have been used by the parties to determine the scope of activities covered by the Treaty as well as to define criteria for the exercise of jurisdiction over the exercises of their legal and natural persons. Gautier (2015) has observed that jurisdiction connection available to States in exercise of their jurisdiction, relate broadly to the nationality or territoriality principles. That said, there is probably no need—and no desire among the States Parties—to develop a uniform approach regarding the exercise of jurisdiction. Gautier (op cit), optimistically concludes that the Antarctic system has been able to evolve whenever faced with new challenges. The issues of jurisdiction too will be solved in the same manner.

4 Illegal, Unregulated and Unreported Fishing (Iuu) Fishing

The problem of IUU fishing and a specific resource allocation issue in the Southern Ocean needs fresh examination in view of state practice as well as treaty law. The IUU is defined as fishing that ignores the provisions of the law of the sea or regulations and arrangement from responsible management bodies (Miller et al. 2010). Freedom to fish is one of the important traditional freedoms of the high seas. However, with the implementation of the UNCLOS and reducing maritime spaces, this freedom has become increasingly fettered. The concept developed in CCAMLR to describe fishing that took place outside the relevant conservation measures set up to protect and manage fish stocks in the CCAMLR region. Concerns raised in CCAMLR meetings led to the matter being raised at the Committee on Fisheries of the Food and Agriculture Organisation (FAO) and in turn to development of the international plan of action to prevent, deter and eliminate IUU fishing.

Initiatives to tackle IUU fishing are good examples of congruity, but it is worth noting that the initiative from within the ATS builds upon the UNCLOS and together with complementary actions outside the ATS, has created a new regime for high sea fisheries. Monitoring remains an important concern, however, state practice of China shows, while it has fishing quota from CCAMLR to fish for krill, that its scientists are assisting in research efforts to monitor fish stocks. US government remains seriously concerned about the need of preventing and discouraging ill fishing vessels within the CCAMLR area. Ross Sea and southern ocean fishing by New Zealand fishing companies brings approximately USD 15 million per annum into the New Zealand's economy. The depletion of fish in these areas has a flow-on effect on the availability of fish in New Zealand's adjacent EEZ. Hence the level of fish stocks in the Ross Sea and Southern Ocean and the rate of IUU is of direct interest to New Zealand. New Zealand is a strong advocate of CCAMLR which manages and monitors fish stocks in the Antarctic seas. South Korea too is very interested in exploiting polar marine products as apparently polar marine products surpass the total production of marine products from all the other waters in the world. South Korea believes that krill is a food resource that will not run out and expect a great economic effect once South Korea finishes establishing the infrastructure on which to begin fishing in earnest.

5 Environment and Climate Change

The landmark Madrid Protocol, signed by Treaty members in 1991 and entering into force in 1998, was a direct result of this initiative. It reaffirms the status of Antarctica as an area reserved exclusively for peaceful purposes, including in particular scientific research, and sets forth a comprehensive, legally binding system of environmental protection applicable to all human activities. The Protocol prohibits all activities relating to mineral resources in Antarctica, except for scientific research. The Protocol also commits Parties to environmental impact assessment procedures for proposed activities, both governmental and private (Comba 2011). The Madrid Protocol serves as a framework instrument that sets out broad principles to protect Antarctica's environment. The main objective of the Protocol is the comprehensive protection of the Antarctic environment and dependent and associated ecosystems. What is the state practice? South Korean scientists have been assessing the greenhouse gas disposal capacity of the Southern Ocean to see if it could absorb still more.

Even for climate change, Antarctica is important to India as the boundary between the Southern Ocean and Indian Ocean is becoming increasingly blurred. It is important to note that New Zealand literature shows that melting of the poles seems reality and present a threat. In fact, climate change in Antarctica will have a direct and immediate effect on its ecosystems and over-fishing of Antarctic sea fish stocks is likely to affect fishing in its EEZ.

In this regard, members states need to be prepared for emergency response actions in the area and arbitration arrangements should be made to resolve international disputes regarding the Antarctic environment. Annex VI of the Protocol focuses on the liability for environmental emergencies. This Annex sets rules governing who is liable for preventing and dealing with environmental emergencies arising from scientific research, tourism and other activities in the Antarctic Treaty area, such as logistic (shipping and aircraft) support. The main aim is to prescribe measures regarding holding persons responsible for cleaning up after an environmental emergency, and the legal avenues to respond to disaster.

6 Oil and Mining

Economic interests, among all interests dominate the operation and governance of the Antarctic regime. Antarctica holds huge wealth of petroleum and natural gas. India does have intricate interest in mineral resources diplomacy and it is beyond any doubt that India's future interest, like in Arctic, will be driven by resource geopolitics.

As far as Asian powers' economic and industrial interests are concerned, South Korea gives the highest importance to these interests. South Korea has dramatically increased its investment in scientific programs and is developing fundamental technologies and biotechnology. South Korea believes that big science is a systematic scientific field, where each discipline interacts with and reinforces one another and Antarctica is a laboratory for big science experiment for South Korea. As Brady and Seungryeol (2013) opine, "if successful, big science will give birth to a raft of new industries" and can lead to huge industrial development. Similarly, South Korea has deep interest in pharmaceutical and construction technology related values to boost its medical and construction industry. Ship building related industries such as steel, paint, and interior design business would be helpful for building ice-capable vessels, which South Korea aims to develop. South Korea, among Asian nations, aim for much larger benefits ranging from industry to addressing issues of food shortage, disease, energy by utilizing undeveloped biological resources of Antarctica. One can observe that South Korea has been embodying its strategic polar interests into more concrete economic and political policies unlike any other nations, especially Asian nations. South Korean research and presence in the Polar Regions is accelerating the shift of its economy from manufacturing to technology and knowledge base industries, and helping to enhance its national brand image so that it can boost the exports of its small and medium businesses.

7 Mineral Exploration, Exploitation and Disguised Scientific Activity

Can creation of mineral treaty guarantee that mining or drilling would necessarily go forward? Perhaps not, but such a regime can help in ensuring that the treaty would be available to regulate such activities if they ever become feasible within the treaty area. Currently, under the terms of the consultative procedure, the Consultative Parties may consider and adopt 'measures regarding…preservation and conservation of living resources in Antarctica'. However, developments in this regard have to be seen in the background of the whole issue of territorial claims. The territorial claims and rejection of the same by several states in Antarctica is one of the most important legal challenges. This is precisely the reason for lack of an effective international instrument governing the minerals. It is believed that "as long as an agreed minerals regime has not been established and approved by both groups, mineral exploitation in claimed territory can be carried out only by permission from the claimant state or in defiance of its claimed authority".

In this regard, it is important to examine Australian state practice. Australia opposed successfully in 1989 not to support any planned regime for regulated mining in Antarctica. It proposed that Antarctic Treaty countries negotiate a comprehensive environmental protection regime for the continent. This initiative resulted into signing of the Madrid Protocol in 1991 which entered into force in 1998. The Protocol reaffirms the status of Antarctica as an area reserved exclusively for peaceful purposes, including in particular scientific research, and sets forth a comprehensive, legally binding system of environmental protection applicable to all human activities. The Protocol prohibits all activities relating to mineral resources in Antarctica, except for scientific research. The Protocol also commits Parties to environmental impact assessment procedures for proposed activities, both governmental and private. Although mining is prohibited by the Madrid Protocol, how to regulate collection of, let us say rare and valuable minerals, on exposed rock near, let us say Australian station, without drilling or using explosives?

It is interesting how state practice shows discretion on mineral diplomacy. For example, South Korean scientists are cautious to avoid using terms such as resource development or resource exploration to discuss their projects. However, much of their current research is focused on locating economically exploitable resources and making more accurate maps of Antarctic resource distribution.

In China, all agree that exploitation of resources in Antarctica is only a matter of time and that China should prepare itself. A section on the website managed by the Polar Research Institute of China to publicize Chinese polar science features a series of maps outlining Antarctic resources, including oil reserves off the coast of Antarctica. It does not breach the treaty to create such a map, but it does send a signal of interest which other states avoid. In 2007 china accused the UK of using law of the sea to gain non-sovereign related rights in Antarctica. The China Arctic and Antarctica Administration's Antarctic oil map and other resource-related maps were produced as part of a wider government sponsored research project on china's marine resources.

8 Governance

Currently, we all know that Antarctic research is being led by individual country's national programs (Duyck 2011). International participation in Antarctic research could be greatly enhanced if there were scientific stations managed under the banner of a collective supranational frame such as the ATS. This is interesting but big challenge. ATS would certainly like to take up this role. Let us examine the state practice in this regard. New Zealand considers that Antarctica is an intrinsic part of their national history and heritage. Collective action in Antarctica would not always be completely sufficient to protect its national interests. Accordingly, New Zealand puts considerable efforts in gaining leadership positions within the ATS. France, taking both political and scientific consideration into account is pursuing several main cooperative strategies most notably the negotiation of both formal and informal agreements and the development of bilateral and multilateral partnerships, centered around multi disciplined scientific collaboration, environment management, resource pooling and logistics sharing the development of a coordinated European polar strategy. In fact, France has developed a coherent and concerted three-tiered strategy—domestic, EU and multilateral level to ensure protection and promotion of its interests through the governance system of ATS. As far as giving serious interest to the governance regime is concerned, no country matches French efforts. For example, no country has appointed a former Prime Minister for international negotiations relating to arctic and Antarctica regions, France appointed its former Prime Minister Mr Rocard in 2009 for this purpose specifically.

As the complexity and interdependent nature of issues concerning the region widen in scope and magnitude, many member states especially those with long-standing and well-established interests are increasingly engaging in cooperative efforts at a variety of levels.

9 UNCLOS, Continental Shelf and Antarctica

There is undoubtedly intrinsic link between the UNCLOS, the regime of continental shelf and the current and future governance of Antarctica. Several states have submitted their data to CLCS showing claims over Antarctica although such claims have been deferred and excluded from the consideration by CLCS (Hernandez-Salas 2015). Despite the submission of claims, the status quo is the only solution. Article 76 of the UNCLOS provides the procedural conduit by which coastal states can gain legitimacy for making sovereignty claims over vast areas of submarine continental shelf areas offshore their coasts—areas that might hold enormous reserves for hydrocarbon resources. Australia was the first to lodge its data on its claims to an extended continental shelf but requested that the CLCS not examine the data associated with the AAT. New Zealand took advantage of an alternative practice available to all parties to the UNCLOS in relation to the delimitation of its extended continental shelf by lodging a partial submission to CLCS. For example, it did not include data on New Zealand Ross dependency claims but the New Zealand government stated that it reserved the right to do so in future. The New Zealand position was that this meant that they were respecting the requirements of the treaty, at the same time as protecting their rights in the future to make a claim to the Antarctic seabed surrounding the Ross Dependency.

An interesting example of international politics affecting international legal regime of ATS related to the intersection between the UNCLOS and the Antarctic Treaty. As is well recognized, the UNCLOS does not directly address Antarctica but covers the maritime areas within the treaty area. A number of commentators have noted that the Antarctica was not mentioned in the deliberations of the 3rd UN Conference on Law of the Sea, while the Antarctic Treaty does not derogate from any rights under the law of the sea.

While all claimant states have declared maritime zones of their Antarctic territories, these zones vary in type and on the basis of their establishment. This is another example of bifocalism, where enforcement of laws related to these zones is made against nationals and vessels flagged to the respective claimant state, not against foreign nations or vessels. Commentators have noted that a number of questions arise in relation to maritime zones, including the question of determining baselines. Some have even questioned whether coastal states exist in Antarctica given the provisions of the Antarctic treaty, and the prohibitions on extending or altering a claimant's territory under the provisions of article IV of the treaty.

10 Tourism

There are issues like how to make Antarctica more accessible to a greater number of nations, how to deal with non-state actors in Antarctica, how to regulate tourism and how to deal with the challenge of bio-prospecting. Ship-borne tourists visiting the South Polar Region has been steadily though not exponentially increasing. As far as tourism is concerned, USA assumes a major interest in Antarctica because of its citizens visiting Antarctica in more numbers than any other country and Antarctic tourist expeditions are subject to US regulations because they are organized in or proceed from the USA as determined by the US State Department, through the Oceans and International Environmental and Scientific Affairs (OES). The US National Environmental Protection Agency (NEPA) reviews environmental impact assessment submitted by such operators and National Science Foundation issues permits to tour operators related to waste management on their voyages. As far as internal governance of domestic policy on Antarctica is concerned, US provides an excellent example. New Zealand Antarctica tourism includes sea tours and New Zealand based companies must include a New Zealand government observer to ensure that the proper environmental protocols are followed. The environmental risks posed by rapidly increasing Antarctic tourism and the politics of regulating this activity is an important legal challenge. Tourism is to be regulated —they pose challenges such as sovereignty, resource protection, governance, regime compliance and problem of how to manage the activities of commercial operators based in countries which are non-signatories. In this regard, an idea of a new multilateral tourism convention on Antarctic level may not be out of place.

11 Strategic Interests

The international peace and stability established in Antarctica by the treaty parties remains indispensable for successful pursuit by the US of its core interests in the region and has become an important objective in its own right. US believes that it may seem ironic to apply a national security yardstick, with its military connotations, to an area that is effectively a demilitarized and nuclear-free zone, however, the potential for international discord and conflict over Antarctic territorial claims which would exist without the treaty today seems greater than was the case in 1959. USA considers itself to be a vital player in Antarctic affairs such that it substantially benefits from effective operation of the treaty. US presence accords a decisive role in the treaty activity based decision system and in maintaining the political and legal balance that makes the treaty function effectively.

12 Principle of Common Heritage of Mankind

India continues to perceive Antarctica as a unique but common heritage of mankind, largely within the framework of the ATS but apparently, neither India nor China has presented any comprehensive working paper on this important issue at the ATCMs. Yet, the 147 sovereign states that are not signatories to the Antarctic Treaty have a right to benefit from what can be gained from Antarctica. While discussing the CHM, it is important to note that New Zealand and USA are proposing that the Ross Sea become a marine protected area though traditional fishing areas will be excluded. South Korea has not yet stated its views on this new initiative. New Zealand regards Antarctica as a national asset which gives it much needed leverage in its dealing with other nations. Chinese scholars tend to focus on preserving the environmental heritage of Antarctica.

13 Liability Regime

Non-claimant states are increasingly affecting decision-making process on ATS regime. Similarly, exercising jurisdiction over non-nationals for criminal, liability of contractual and torts need to be resolved (Wolfrum and Wolf 2008). As the treaty law is silent, the reliance can be placed to an extent to emerging case-law which is very little. Indian position appears to be quite silent, as there are no specific references during the ATS meetings or in governmental public documents. The available literature shows that US national law does apply to the Antarctic, in particular to nationals of treaty parties visiting the continent and to the management of national research stations on and around the continent. Both civil and criminal law as they relate to activities by US citizens, both in research stations and working on US vessels or in fields camps in the Southern Polar Region. What are the major decisions of the US courts in this regard? Whether the Federal Tort Claims Act (FTCA) applies to tortuous acts or omissions occurring in Antarctica, a sovereign less region without civil tort law of its own? (Sandra Jean Smith V United States, 507 U.S. 197) The FTCA does not apply to claims arising in Antarctica. Secondly, whether Antarctica is a "foreign country" under the Fair Labours Standard act, 1938? (Smith V Raytheon, 297 F. Supp. 2d 399), US court concluded that Antarctica is a "foreign country" within the meaning of section 213(f). There can be no disagreement over the proposition that Antarctica is "foreign" to the United States. And, the Supreme Court has broadly defined the word "country" within the term "foreign country" to mean "[a] region or tract of land." Antarctica is certainly "[a] region or tract of land." It, therefore, is not necessary to look any further than the ordinary, commonsense meaning of the language of section 213(f) to reach the conclusion that Antarctica is a "foreign country" within the meaning of that section. Thirdly, whether Antarctica is a "foreign country" within the meaning of income tax provisions (Martin v Commissioner, 50 T.C.59; T.C. 1968). The term "foreign country" means territory under the sovereignty of a government other than that of the United States and includes the air space over such territory. It does not include a possession or Territory of the United States. Antarctica plainly does not fall within these terms. The parties have stipulated that "the Department of State does not consider the Antarctica region to be under the sovereignty of any government," and it is well established that in matters of foreign affairs generally and in determinations of sovereignty in particular, the courts must accept the position taken by the executive branch. Moreover, even apart from the claims of some nations to portions of Antarctica, the area which may be referred to as the United States sector (which contains Byrd Station) does not appear to be claimed by any government. Accordingly, it is clear that neither Antarctica as a whole, nor that portion containing Byrd Station, is a "foreign country" within the meaning of the regulations. Fourth, whether a federal agency may decide to take actions significantly affecting the human environment in Antarctica without complying with NEPA and without being subject to judicial review? It cannot be seriously suggested that the United States lacks some real measure of legislative control over Antarctica. The United States controls all air transportation to its bases in Antarctica and cooperates in search and rescue operations there (Environmental Defense Fund v. Massey, No. 91-5278, US Courts of Appeal, District of Columbia (Circuit (1993). 986 F.2d 528). Moreover, the United States has exclusive legislative control over McMurdo Station and the other research installations established there by the United States Antarctica Program. This legislative control, taken together with the status of Antarctica as a 'sovereign-less' continent, compels the conclusion that the presumption against extraterritoriality is particularly inappropriate under the circumstances presented in this case. As stated aptly by a State Department official in congressional testimony shortly following the enactment of NEPA, "application of NEPA to actions occurring outside the jurisdiction of any State, including the United States, would not conflict with the primary purpose underlying this venerable rule of interpretation—to avoid ill-will and conflict between nations arising out of one nation's encroachments upon another's sovereignty. There are at least three general areas: The high seas, outer space, and Antarctica. Applying the presumption against extraterritoriality here would result in a federal agency being allowed to undertake actions significantly affecting the human environment in Antarctica, an area over which the United States has substantial interest and authority, without ever being held accountable for its failure to comply with the decision making procedures instituted by Congress-even though such accountability, if it was enforced, would result in no conflict with foreign law or threat to foreign policy". Thus, the U.S. foreign policy on global commons, including Antarctica, do not present the challenges inherent in relations between sovereign nations. Thus, in a 'sovereign-less' region like Antarctica, where the United States has exercised a great measure of legislative control, the presumption against extraterritoriality has little relevance and a dubious basis for its application.

14 Dispute Settlement

ATS has remained highly successful and effective when assessed against its core provisions of managing disputes over territorial claims and sovereignty in Antarctica and ensuring that Antarctica should be used exclusively for peaceful purposes. Articles I, II, and IV and the statement in the preamble to the Treaty that Antarctica shall not become the scene or object of international discord, provide the basis for political interaction within the ATS.

There is very little oversight of states behaviour there—technically any state can inspect other countries Antarctica bases but few devote the resources to do this with any seriousness. Moreover, most states ignore the legal requirement to make fully public all their Antarctic activities and capabilities, while some ATCPs are engaging in very low level research.

15 State Practice in General—Concluding Remarks

State practice in Antarctica shows a strong fragmentation of international law which can lead to conclusion that the ATS is not coping well with the needs of the emerging world order and its power imbalance certainly reflects the realities of that order. Secondly, the desire of the most nations engaged with Antarctica is the desire to maintain access to current and future resources there. The ATS in its present form suits the needs of the established Antarctic players only, but still appears unable to meet the needs of less established states. And yet for all its failings, no other instrument of governance is available to deal with all the issues involved in governing the Antarctic continent and Southern Ocean (Huber 2008).

16 Analysis of Evolving Position of India—Concluding Remarks

What the above examination of state practice and functioning of the treaty regime lead us to draw conclusions for India? India remains committed to scientific research and technical cooperation in the Polar Regions. Indira Gandhi, the then Prime Minister of India, in 1983, had said "The Indian Ocean links India to Antarctica. The entire area is of deep interest to us, and ocean studies are of vital importance" President Mukherjee, the then Defence/Foreign Minister in 2007 at A. K. Chatterjee Memorial lecture mentioned that "not only is Antarctica vitally important for the environment, it is a treasure house of potential mineral resources, including petroleum. It is an enormous marine storehouse for the foundation of the human food chain, thanks to its abundant holdings of krill. Antarctica determines in significant measure the Indian monsoon—upon which our agriculture and hence our economy depends. …it would by now, be obvious that the primary area of Indian maritime interests range from the Persian gulf in the north to Antarctica in the south…" (Chaturvedi 2013). India continues to believe that Antarctica is a common heritage of mankind and the foremost symbol of peaceful use and cooperation that needs to be protected for posterity. What emerges is, like our oceans, a grand strategy is conspicuous by its absence which could provide India multiple strands a cohesive form, consistency and orientation. As said earlier, New Zealand considers Antarctica as its national asset. India should work towards ensuring that Antarctica is considered as global asset. India is a long way to go to strategic culture whereby institutionalisation of foreign policy making can become possible. As per Chaturvedi (2013) "the link between the absence of institutionalisation of India's foreign policy and the absence of coherent Antarctic policy, even after three decades of substantial scientific engagement with the southern polar region, calls for some serious thinking".

The wealth of state practice suggests that India ought to adopt and follow, to suit its requirements, multiple interests in Antarctic just moving beyond polar science. While ratifying the Convention, India believed that it would be in a much better position to enhance its analytical capabilities. Furthermore, India being then leading proponent of NAM articulated its interest in terms of NAM as it believed that it will be in a position to effectively projects its own views as well as the views of the nonaligned countries and that it would be able to participate in the ongoing discussions on the resources and ensure that any regime that might be set up there would be in harmony with its overall policies and objectives. While these were then prevailing just 'reasons for the government to ratify' the instrument, Indian practice shows that it has lacked credibility in terms of its commitment to codify its interests in national legal instruments. The continuous absence of legal instruments has created an atmosphere of adhocism which can create situation of unaccountability and which furthermore keep institutions and individuals in disarray. Speaking through Mrs Gandhi, the then Prime Minister, India proclaimed that India will not do nor allow others to take any steps which would be considered detrimental to the interests of the developing countries. There is a huge gap between rhetoric of 1970s and newly enlightened national interest of the 21st century. China which is late comer has taken measures and scaled up operations unlike any other nations, risking even its image and the continuous rise is speedily displacing the interests of existing powers including India.

Indian legal framework and policies should address intricate issues of scientific research including intellectual property, technology transfer, among others. India has signed various maritime conventions and it is known that there is an extensive array of instruments that can apply to Antarctica and the Southern Ocean, such as MARPOL 73/78, CITES, UNFCC, ICRW and SOLAS.

Chaturvedi (op.cit) in his concluding note says that "Asia's rise in the international system, and its implications for the Antarctic, can legitimately be approached and analyzed in the light of the forthcoming era of shortages in five vital areas governing economic activity; food, commodities, energy/oil, water and clean environment". The law and policy of India should cover the achievements of objectives through such a framework.

Nations' efforts to affect international law for Antarctica will change according to their interests and history. Although India ought to have multiple interests, the current regime of limited interests enable very little input from Indian establishment to contribute to the international legal instruments concerning Antarctica. In France, a rich school of literature already exists on issues relating to Antarctic law and governance within international environmental law and politics. Numerous studies have focused on issues relating to regime formation, levels of state participation, and problem associated with treaty compliance and enforcement.

Although India may have little less to gain from whaling, it is noted that whaling remain quarantined from discussion in the ATS, India should develop its position slowly but steadily. Secondly, IUU fishing poses ongoing challenges to the effectiveness of CCAMLR conservation measures. India shall make its stand clear. Third, biological prospecting, as mentioned earlier, has yet to be addressed in an effective manner. Actions by claimant states in relation to the delimitation of continual shelf have shown the success to date of a nuanced approach to this issue but at the same time it may rekindle the ABC issue, a focal point in the negotiation of the Antarctic treaty.

It is clear that Antarctica is governed in the name of science, and it is through scientific activities there that countries can legitimately be involved in decision making about the continent's governance and management of its resources. India does need a clear Antarctica strategy. The skirmish over Antarctic resources is being waged through scientific, diplomatic and legal measures rather than through force, and mineral exploration is continuing, disguised as scientific activity (French and Scott 2009).

Moving beyond science, there is a strong need to create humanities and social science research, as in China. Chinese social scientists are paying greater attention to researching Antarctica issues. This emphasis is directly encouraged by senior government leaders and the outcomes are now being utilized in china's policy making on Antarctica.

Finally, though Antarctic governance is ruled by consensus there are dominant players, those who have greater physical presence (USA and Russia), those who publish the most policy papers (UK, Australia and New Zealand), those who are the most politically active in ATS committees or on new challenges to Antarctic governance (New Zealand, France, Australia over the issue of mining), those who form informal blocs of like-minded states (Australia, New Zealand, UK, USA and Chile). India does not feature in network of any of these dominant players. The science, including big science (like South Korean state practice) international law, social and political science debates need to go hand in hand to understand international order of Antarctica (Jabour 2012).

Acknowledgements The paper draws substantial facts and figures from three latest scholarly and policy-oriented works in order to understand the historical position of several nations and adapt and predict possible position of India, namely, Anne-Marie Brady's editorial work—The Emerging Politics of Antarctica, (Routledge: 2013); Alan D. Hemmings, Donald Rothwell and Karen Scott (eds.), Antarctic Security in the Twenty-First Century: Legal and Policy Perspectives, (Routledge: 2012) and Erik Molenaar, Alex Oude Elferink and Donald Rothwell (eds.) Marine Scientific Research in the Antarctic Treaty System (Nijhoff: 2013). Richard Rowe, International Law and Diplomacy, 15 Melbourne Journal of International Law 318-29 (2014); Richard Caddell, Legislation and Policy, 16 Environmental Law Review 122–127 (2014).

References

Bates R (2011) Touring the Antarctic: transforming environmental governance in the southern latitudes Asia Pac J Int Environ Law 14(1/2):43–62

Brady A-M (2013) China's Antarctic interests. In: Brady (ed) The emerging politics of Antarctica, Routledge, p 45

Brady A-M, Seungryeol K (2013) Cool Korea: Korea's growing Antarctic interests. In: Anne-Marie Brady (ed) The emerging politics of Antarctica, Routledge, p 75

Chaturvedi S (2013) India and Antarctica: towards post-colonial engagement? In: Anne-Marie Brady (ed) The emerging politics of Antarctica, Routledge, pp 630–692

Comba D (2011) The polar continental shelf challenge: claims and exploitation of mineral sea resources: an Antarctic and Arctic comparative analysis. Yearb Int Environ Law 20:156–187

Davies T (2013) Antarctic resources: balancing the challenges and needs, Defence Academy of the United Kingdom

Duyck S (2011) Drawing lessons for Arctic governance from the Antarctic treaty system, Yearb Polar Law, pp 683–713

Fiametta B, Federica M (2014) The legal regime of Antarctic. In: Attard D (ed) The IMLI Manual on International Maritime Law. Oxford University Press, Oxford

Finn S (2010) Jurisdictional problems in relation to Antarctic mineral resources in political perspectives. In: Francisco OV (ed) Antarctic resources policy: scientific legal and policy issues. Cambridge University Press, Cambridge, pp 317–335

Foster C (2012) International adjudication: standard of review and burden of proof: australia-apples and whaling in the Antarctic. Rev Eur Community Int Environ Law 21(2):80–92

French D, Scott K (2009) International legal implications of climate change for the polar regions: too much, too little, too late? Melbourne. J Int Law, p 631

Gautier P (2015) The exercise of jurisdiction over activities in Antarctica: a new challenge for the Antarctic system. In: Castillo LD (ed) Law of the sea, from grotius to the international tribunal for the law of the sea, liber amicorum hugo caminos. Brill, Nijhoff, pp 192–210
Hernandez-Salas C (2015) Distinguished status quo: the American Antarctic quadrant after submissions to the commission on the limits of the continental shelf. Int J Mar Coast Law 2:285–304
Huber J (2008) Some notes on my personal experience setting up the Antarctic treaty secretariat. In: Tamburelli G (ed) The Antarctic legal system: the protection of the environment of the polar region, Milano, pp 153–161
Jabour J (2012) The Antarctic Treaty system: what's on the horizon? Yearb Polar Law, pp 709–722
Joyner C (2012) Bioprospecting as a Challenge to Antarctic Treaty. In: Donald R (ed) Antarctic security in the twenty-first century: legal and policy perspectives, pp 197–214
Mbengue MM (2015) Between Law and science: a commentary on the whaling in the Antarctic case. In: Questions of International Law, pp 3–12
Miller D, Slicer N, Sabourenkov E (2010) IUU fishing in Antarctic waters: CCAMLR actions and regulations. In: Vidas D (ed) Law, technology and science for oceans in globalisation: IUU fishing, oil pollution, bioprospecting, outer continental shelf. pp 175–196 (Nijhoff)
Pincus R (2015) Diplomacy in ice: energy and the environment in the Arctic and Antarctic. Yale University Press
Scott K (2011) Managing sovereignty and jurisdictional disputes in the antarctic: the next fifty years. Yearb Int Environ Law, pp 3–40
Scott K (2012) Scientific rhetoric and Antarctic security. In: Donald R (ed) Antarctic security in the twenty-first century: legal and policy perspectives, pp 284–306
Scovazzi T (2015) Between Law and Science: some considerations inspired by the whaling in the Antarctic judgment. Questions Int Law, pp 13–30
Tony D (2013) Antarctic resources: balancing the challenges and needs, defense academy of the United Kingdom
Weber M (2008) Australia's extended continental shelf submission: states' right in the context of Antarctic regional governance. Antarct South Ocean Law Policy Occas Pap 12:1–32
Wolfrum R, Wolf S (2008) The Antarctic liability regime. In: Tamburelli G (ed) The Antarctic legal system: the protection of the environment of the polar regions, pp 161–183

Arctic: A Paradox and Antithesis

Uttam K. Sinha

Abstract Due to its resource potential, sensitive ecological location and a common play ground for several nations, Arctic will continue to play a strategic role in the fields of maritime, climate change and international partnership policies. While the alarming predictions of IPCC indicate a future scenario of sea-ice free Arctic, the opening up of new sea routes may redraw the geo-economic map of growth and development in the Arctic. This may call for a balanced appreciation of climate change risks and opportunities of hydrocarbons explorations in the Arctic, notwithstanding emerging financial and practical problems in extraction of hydrocarbons. In spite of the existing UNCLOS laws pertaining to the international seas, a Polar Code may have to be devised for Arctic in line with Antarctic Treaty System for the purpose of management of scientific work and logistics of exploration. Sustainable development, as is generally the principle, must be inclusive of strong environmental goals in Arctic too. Asian countries, as the observers in the Arctic Council have a responsibility to highlight the impact of warming Arctic to rest of the world as teleconnection between Arctic climate and Himalaya-Tibet plateau has scientifically been established.

Keywords Climate change · Common heritage · Sustainable development

1 Introduction

In an interconnected world with interlinked issues, observing the geo-physical changes is critical. The Arctic is witnessing the convergence of the geophysical, the geo-economics and the geo-strategic in strange and dramatic ways making it a paradox and an antithesis. The changing landscape is keeping countries, particularly

U.K. Sinha (✉)
Institute for Defence Studies and Analyses, New Delhi, India
e-mail: uksinha2001@gmail.com

U.K. Sinha
Malaviya Centre for Peace Research, Banaras Hindu University, Varanasi, India

P.S. Goel et al. (eds.), *Science and Geopolitics of The White World*,
DOI 10.1007/978-3-319-57765-4_4

those within the rim and some beyond, honest in their engagement. Inevitably, competition and cooperation will emerge, along with positioning and posturing. Four reasons for increasing attention on the Arctic can be highlighted. First, the Arctic will continue to remain a large geo-strategic tract. Whether it is new resource finds or new emerging transport routes, the Arctic's strategic value will only amplify. Second, since the Arctic is a semi-enclosed ocean surrounded by land, and like all high seas, governed by the laws of the sea (UNCLOS). Understanding the legal regimes that applies for navigation in new shipping routes and those that governs the exploitation of the vast oil and gas resources will be challenging as it will directly confront states' interest. Third, relates to the question of resource finds. The Arctic holds vast untapped gas reserves, potentially large undeveloped oil reserves and mineral wealth, making it the final frontier for energy development. Any resource development will require building massive infrastructure through areas that are ecologically sensitive. The fourth Arctic attention concerns the sea routes. Clearly, the Arctic is becoming accessible to a number of different actors with varied, and not mutually beneficial, agendas. How then does the world navigate through this 'interplay' and move towards a global knowledge commons that includes scientific understanding and learning, ecological protection and sustainable use of resources—what can be described as an 'Arctic governance web'.

2 Science of Climate Change

In March 2013 and then in April 2013 the Intergovernmental Panel on Climate Change (IPCC) released the Working Group 2 and 3. These are summary for policy makers. Briefly, WG2, in no uncertain terms, warned of the increases in frequency of extreme weather events from the impact of climate change. WG3, on the other hand, focused on solutions to curb carbon emissions by assessing mitigation options in different economic sectors. Without additional mitigation efforts, the report says that the world may be headed to a 3.7–4.8 °C temperature increase by the end of the century. Worrying was the fact that in spite of great attention to climate change mitigation policies worldwide, the annual greenhouse gas (GHG) emissions grew on average 1.0 billion tonne of GHG per year from 2000 to 2010.

Some of these impacts captured in the Summary for Policy Makers (SPM) are:

- *Warming of the climate system is unequivocal, and since the 1950s, many of the observed changes are unprecedented over decades to millennia. The atmosphere and ocean have warmed, the amounts of snow and ice have diminished, sea level has risen, and the concentrations of greenhouse gases have increased.*
- *Each of the last three decades has been successively warmer at the Earth's surface than any preceding decade since 1850. In the Northern Hemisphere, 1983–2012 was likely the warmest 30-year period of the last 1400 years.*

- *Ocean warming dominates the increase in energy stored in the climate system, accounting for more than 90% of the energy accumulated between 1971 and 2010.*
- *Over the last two decades, the Greenland and Antarctic ice sheets have been losing mass, glaciers have continued to shrink almost worldwide, and Arctic sea ice and Northern Hemisphere spring snow cover have continued to decrease in extent.*
- *The rate of sea level rise since the mid-19th century has been larger than the mean rate during the previous two millennia (high confidence). Over the period 1901–2010, global mean sea level rose by 0.19* [*0.17–0.21*] *m.* (IPCC 2013).

3 Arctic Attention

The Arctic given the geophysical changes is primarily a destination for scientific research expeditions and observation of the changes in the region—what can be referred to as an 'ecological' perspective. But any serious look at the Arctic cannot ignore the fact that because of the meltdown the Arctic has given way to an extremely active geopolitical space. An immediate impression is that geo-economics readily converges with the opportunities that the geophysical changes in the Arctic present. High rate of industrial development causes the strong growth of demand on raw materials in particular hydrocarbons. Can this be interpreted as redrawing the geo-economic map of growth and development in the Arctic? Or a balanced ecological appreciation is required in terms climate change risks and vulnerability than the real politic imperatives to explore the opportunities of hydrocarbons explorations in the Arctic? Four reasons can be identified for Arctic attention.

The first of these attentions relates to the geopolitical considerations, which is dynamic and changeable and in many sense unique in the Arctic. Each country irrespective of its size or political influence has an impact in the Arctic pertaining to the issue at stake. Some of these issues are settled multilaterally while others are resolved bilaterally. The Arctic region, with advanced and highly developed countries, simultaneously witnesses clashes of interest and collaborations and where allies can turn adversaries and vice versa. Arctic's strategic value will only amplify, whether it is resource exploration and extraction or new emerging transport routes. This will, in spite the fact that tensions have remained historically low, create a new set of political disagreement and resultant high politics. The race for resources inevitably leads to shove and push, making the Arctic potentially an area of competition and contestation. On the other hand, opening of shipping routes are likely to foster new cooperation and stimulate regimes and mechanisms. The fluctuating geopolitical changes will impact whether the region will experience competition or cooperation or simultaneously both. The Arctic's political temperature may heat up in different ways. The immediate reasons could be the discussion on 'who' shall extract the oil when the ice thins and possibly disappears?

'How' will the new marine delimitation lines be drawn? 'Who' will control the new sea passage? And maybe at some stage a bigger question on 'who' owns the Arctic will be raised (Sinha 2013)?

The second Arctic attention is the legal consideration. It must be remembered that the Arctic Ocean is a semi-enclosed ocean surrounded by land and like all high seas is governed by the United Nations Convention on the Law of the Sea (UNCLOS). The Antarctica, a geographical contrast, is a landmass surrounded by an ocean. UNCLOS which was adopted in 1982 and came into force in 1994, did not envisage a special regime for the Arctic—the Arctic Ocean was no different from any other oceans. However, under Article 234 it gave the Arctic coastal states special regulatory and enforcement rights to reduce and control vessel source marine pollution within the limits of the exclusive economic zone (EEZ). So while there are norms and regulations, differences may emerge on the interpretation of the existing laws because of the geophysical changes in the Arctic. UNCLOS provides the universal regime for all matters relating to ocean affairs and the law of the sea. It serves as the basis for the development of regional and national ocean policies, as well as the development of related regional and international instruments (IDSA 2013). To be actively involved in the Arctic it is vital to understand what exactly is the legal regime under the Convention and what issues are indeed critical and whether the Convention provides an adequate framework for dealing with those critical issues especially the legal regime that applies for navigation in new shipping routes and those that governs the exploitation of the vast oil and gas resources (IDSA 2013).

All Arctic states with the exception of the USA are parties to UNCLOS and all including the USA agree that the legal regime contained in the UNCLOS applies to the Arctic as well. However, while laws exist, they regularly clash with sovereignty. The Arctic too may witness claims and counter-claims before being settled. India with its long polar research experience, active involvement with the law of the sea negotiations for over fifty years and experience in deep sea exploration is uniquely placed to contribute effectively on several scientific projects management as well as logistics of expeditions.

The third Arctic attention refers to the resource finds. The Arctic, it is said, holds the largest remaining untapped gas reserves and undeveloped oil reserves. It is dubbed as the final frontier for energy development. Much of these potential reserves lie offshore, in the Arctic's shallow shelf. One often gets the impression that the Arctic's oil and gas finds will be the answer to the world's energy thirst. It is though forgotten that many known reserves are not exploited because of their inaccessibility—short productive period and low temperatures. Any oil and gas development will require building massive infrastructure through areas that are ecologically sensitive. Not surprisingly, Royal Dutch Shell's exploration and drilling efforts during the summer season of 2012 in the Chukchi Sea fell behind expectations. Likewise, Cairn Energy's high investments in exploratory wells in Greenland's coast have made no commercial discovery. The other resources in the Arctic are the vast minerals. The Arctic region of Russia is probably the most developed and has vast deposits of nickel, copper, coal, gold, uranium, tungsten, and diamonds.

It goes without saying that oil and gas extraction in the Arctic is inherently dependent on commercial profitability. With technical difficulties brought about by location challenges, weather hazards and unresolved maritime boundary disputes, the cost of Arctic oil production, according to IEA, will be very high. On the gas front, following a sudden boom in the US domestic shale gas production and with price levels falling rapidly, the natural gas from the High North will not probably be competitive also keeping in fact the difficulties in extraction. There is also strong environmental concerns vis-à-vis the development of Arctic resources. The 1989 Exxon Valdez oil spill in Alaska and the more recent Deepwater Horizon accident in the Gulf of Mexico in 2010 are reminders of the potential hazards and ecological consequences of drilling activities. In September 2015, the Royal Dutch Shell ceased its oil exploration in the Chukchi Sea in Alaska Arctic after disappointing drilling results. Exxon Mobil too has shelved its plan to drill in Russia Arctic and likewise Chevron is not too excited to carry forward its oil exploration in the Arctic. Oil and gas companies operate on profit and it is not yet clear what that profit margin is for the companies to venture dangerously out in the Arctic. The initial euphoria over the oil and gas finds have given way to ground reality to the point that the Arctic new status as the 'new energy province' is being questioned. Striking a balance between economic interests and lowering environmental risks will be a challenge but crucial to the resource management and governance. Given the current situation it is unlikely that Arctic will emerge as a major contributor to global energy resources (Østhagen 2013).

The fourth Arctic attention concerns the sea routes. As expressed earlier, given the various physical difficulties and the global economic downturn, extraction of oil and gas will be low key thus making the Arctic more an active shipping route rather than an oil and gas production zone. With the Arctic meltdown, while new shipping routes will open up, the rights of states for various types of passage (innocent, transit, archipelagic or free passage) are already set out in UNCLOS. However, practical modalities and implementations have to be worked out. It is in this light that the Polar Code needs to be developed. The Antarctic Treaty System could serve as a model for the Arctic too particularly on how cooperation and coordination can be effectively put in practice, both for scientific work and logistics of expeditions.

4 Arctic Message

With rapid geophysical changes in the Arctic region, the need for actionable Arctic science has never been greater. An increasingly warmer climate is impacting the Arctic ecosystems with loss of sea ice and glaciers, thawing of permafrost, and changing snow patterns. These climatic shifts challenge the people who inhabit the Arctic region and who must adapt to new environmental changes. It is thus important to understand how environmental and societal transitions will affect the

Arctic and the need to translate research findings into practical information that can help guide policy decisions.

Some of the messages that the Arctic change is sending across are:

- With reduced ice and snow the Arctic societies are also changing, especially in the political realm as indigenous people achieve greater autonomy in some regions.
- Reduced ice cover now allows research all year around, in new fields, and new geographical areas. Research and scientific investigations will be paramount as the ice barriers diminish. Research is essential to understand the drivers of change, their implications, and options for response.
- The Arctic is not desolate and remote. It is inextricably connected to the rest of the world through climate and meteorology and even by animal migrations and societal interactions.
- Enhanced research cooperation and collaboration between varied institutions, across disciplines and among nations will be needed to answer the fast changing geo-physical landscape.
- New technologies and advancement in sensors and satellites will help allow new approaches to research in many fields especially the ocean and atmosphere. The Arctic research will require sufficient human capacity, including scientists trained in the necessary fields who are capable of interdisciplinary collaboration.
- Collaboration will be critical not only among scientific disciplines and between scientists but also with decision makers to better understand the Arctic and how scientific results will help produce decision outcomes.

5 Arctic: New Ways of Thinking

5.1 Multi-lateral Development Bank

As explained, the Arctic has an array of complex real-world problems. To overcome challenges from time to time states or a coalition of states have thought about innovative ways and come up with ingenious solutions. To recall, the Multi-lateral Development Bank (MBD) model was conceived to deal with the daunting task of reconstructing Europe and Asia in the post WWII. Similarly such institutional approach can be considered to help fund international development in the Arctic. Increasingly, the Arctic will demand a governance structure that would embrace the interest of all developed countries, while respecting the economic sovereignty and environmental regulations and the interests of their northern inhabitants (Gill and Svigny 2015). Of course there are many unsettled questions regarding the role of MBD in the Arctic but it does merit a dialogue between multilateral financial experts and Arctic specialists regarding the advantages of MDBs and whether it can help meet the Arctic region massive requirement, in the coming years, for new public infrastructure and private investment.

With the continued thaw of the Arctic ice cap, new polar shipping routes such as the Northern Sea Route and the Northwest Passage are receiving greater attention. Among the most important challenges are need for responsible resource development, safe Arctic shipping and sustainable circumpolar communities. All these specific areas will require new investments in critical infrastructure like improves road and rail networks, deep water ports, airports and runways (Gill and Svigny 2015). The lack of this infrastructure inhibits resource development the lack of which has also slowed the growth of sustainable Northern communities.

The MDB are now far more experienced than the days of operation in the post-WWII and have through the decades gained wide experience in working with governments and the private sector, especially in the extractive industries. While, of course resource development in the North is dominated by mining activity but there is great scope in areas such as stronger governance structures and improved monitoring of environmental and social performance.

5.2 Developing Sustainable Development Goals in the Arctic

There is a tendency to view environmental protection and sustainable development as mutually exclusive. In fact, sustainable development includes strong environmental goals. To recall, one of the main outcomes of the Rio +20 Conference was the agreement by member States to develop a set of Sustainable Development Goals (SDGs), which will build upon the Millennium Development Goals and converge with the post-2015 development agenda. The agreement stated: "*inclusive and transparent intergovernmental process open to all stakeholders, with a view to developing global sustainable development goals to be agreed by the General Assembly*". In the Rio +20 outcome document, member States agreed that sustainable development goals (SDGs) must be:

- Consistent with international law
- Address and incorporate in a balanced way all three dimensions of sustainable development and their inter linkages
- Include active involvement of all relevant stakeholders, as appropriate, in the process

It must equally be:

- Useful for pursuing focused and coherent action on sustainable development
- Serve as a driver for implementation and mainstreaming of sustainable development in the UN system as a whole
- Address and be focused on priority areas for the achievement of sustainable development.

While the challenge facing the Arctic is to continue the environmental protection, it equally recognises that it must link it more closely to sustainable development. Mary Simon, Canada's former Ambassador for Circumpolar Affairs, had remarked that the Arctic Council must not make the mistake of seeing environmental protection and sustainable development as distinct, as the AEPS had done, but that sustainable development must have strong environmental goals.

Therefore, most of the goals of sustainable development as mentioned above are important for the Arctic region. Many scholars for example, Oran Young has emphasised that sustainable development should be the overarching framework for the Arctic Council as it sets out to chart new developments in international Arctic cooperation. Young has further recommended that subsistence preference, co-management, and the development of environmentally-appropriate technologies and practices should be some of the guiding principles for the Council's work on sustainable development.

Sustainable development is a priority area for the Arctic Council, which follows closely the 1987 Brundtland Commission definition as development which meets the needs of the present without compromising the ability of future generations to meet their needs. Interestingly, Canada, in its role as first chair of the Council, defined sustainable development as "development which seeks human well-being through an equitable and democratic utilisation of society's resources, while preserving cultural distinctiveness and the natural environment for future generations".

5.3 *Arctic and Asia: A Scientific Enterprise*

The climate looms large in discussions on the Arctic and the Himalaya-Tibet in Asia, where the impact of global warming will have significant ramifications in terms of competition for resources and managing the fragile ecosystem. The Arctic and the Himalaya-Tibet are possibly the most environmentally strategic areas of the world. There are immediate areas of convergence on issues such as resource use, sustainability and global governance both in the Arctic and Himalaya-Tibet. The Arctic is one of the original poles, while Himalaya/Tibet has recently become to be regarded as the 'Third Pole'. In both places protecting the ecology should be high priority. The ecological footprint in the Arctic is heavy, with emissions accounting for up to 45% of black carbon and 25% of all mercury (Overseas Territory Review 2011). The glaciers of the Himalaya-Tibetan Plateau contain one of the largest reservoirs of snow and ice outside the Polar Regions and are a source of many major Asian rivers. Retreating glaciers, melting permafrost and degrading ecosystems with monsoon variability are the consequences of ongoing regional and global climate warming. Consequently, shrinking glaciers will result in the decrease of water runoff in the long-term. In the short-term earlier water runoff from glaciers when combined with seasonal rains can result in flood conditions.

The Arctic and the Himalaya-Tibet, though geographically distant, are interconnected and share similar concerns. Fluctuations in the glaciers of the

Himalaya-Tibetan Plateau are the result of environmental changes on a local, regional and global scale. On the other hand, the melting of the Arctic is likely to result in sea-level rise and alter the stable patterns of ocean currents resulting in unpredictable weather cycles. Scientists reason that the Himalaya-Tibetan Plateau is not only a key component of Asian monsoon evolution but that the fluctuations on the Tibetan glaciers have a significant impact on the climate system in the Northern Hemisphere and on the entire earth on various temporal and spatial scales (Hasnain 1967). Studies have also shown a significant co-relationship between the Arctic Oscillation (AO) and the autumn-winter snow depth on the Tibetan Plateau (Lü et al. 2008). Scientists believe that the AO is causally related to weather patterns in areas thousands of miles away, including many of the major population centres of Europe and North America. NASA climatologist James Hansen explains the mechanism by which the AO affects weather at points so distant from the Arctic: 'When the AO index is positive, surface pressure is low in the polar region. This helps the middle latitude jet stream to blow strongly and consistently from west to east, thus keeping cold Arctic air locked in the polar region. When the AO index is negative, there tends to be high pressure in the polar region, and greater movement of frigid polar air into middle latitudes (Hansen et al. 2009).

Before the late 1970s when the AO was in its inter-decadal negative phase, the snow depth over the Tibetan Plateau increased in autumn and then decreased in the following winter. Now the AO has been in a positive phase since the early 1980s, and consequently snow depth has decreased. Furthermore, sediments taken from the bottom of Kiang Lake on the Tibetan Plateau suggest that changes in wind patterns, which are clearly caused by global warming, are making the area dustier (Science Now 2011). According to the American Geophysical Union, this trend could accelerate the melting of crucial glaciers in the Himalayas and affect already imperilled water supplies.

The increase in dust particles in the Tibetan plateau was at one time attributed to over-grazing and increased activity by local people. Scientific observation has now revealed that dusty periods coincide with the AO being in a 'positive phase'. As a result of this positive phase, the Tibetan plateau is exposed to stronger winds in the summer. The link between dust levels and the AO, while not exact, does indicate that a dustier atmosphere can accelerate the melting of the glaciers in the Himalayas. Common science tells us that as dusts settles on white ice, it makes it darker thus absorbing radiation and accelerating melting. Dust also warms the air above, enhancing monsoon circulation patterns which could affect rain and alter rainfall patterns.

As the Arctic melt raises the sea level, the Tibetan glacier melt will increase the flow to many rivers, from the Yangtze, which irrigates more than half of China's arable land, to the Indus river system, which is critical to the agricultural heartlands of India and Pakistan. Ongoing studies suggest that 40% of the plateau's glaciers could disappear by 2050. Studies also indicate that full-scale glacier shrinkage is inevitable and will lead to ecological catastrophe.

The impact of warming on the Tibetan glaciers and its direct relation to river flows creates an opportunity for the down river countries in South and Southeast Asia to raise common concerns and draw China into a regional dialogue and joint study on the climate change impact on the glaciers and precipitation patterns. In the Arctic Council, China and India as key observers can initiate a study on the snow, water, ice and permafrost of the two regions. The findings of such studies will help both the countries to prepare preventive polices in the region. Down River countries dependent on the rivers from Tibet should also advocate the establishment of a new ecological regime for the protection and sharing of the Tibetan Plateau. In the case of the Arctic as well, similar questions can be raised. As has been noted, 'should five countries, which, as an accident of geography, form the Arctic rim, have the right to play with the world's ecological future in pursuit of their economic interests (Saran 2011)?

There are also common concerns and changing realities for both the Arctic and Tibet that need to be addressed, particularly with regard to whether the resources of the Arctic (oil and gas) and Tibet (freshwater) can be regarded as 'global commons' (Gautam 2011), or as the 'common heritage of mankind'. While no two issues can be the same and one should be cautious of drawing parallels. Moreover, many states would contest the principle of 'global commons' or 'common heritage' based on sovereignty and territorial jurisdiction. In the latter part of the twentieth century, the term 'commons' has expanded to include intangible resources such as the internet, open-source software, and many aspects of culture. The term 'global commons' is more recent and has several meanings: those resources that are shared by all of humanity, such as the sky, the oceans, or even the planet itself; the sum of various local and regional commons across the world; and a philosophical position suggesting that humankind has both a right and a responsibility to steward the wise use of the earth for all living species, as well as for future generations (UNITAR 2013). The concept of the common heritage of mankind was first articulated in 1970, when the UN General Assembly adopted a Declaration of Principles governing the seabed and ocean floor. Now this concept includes outer space, the legal status of lunar minerals, geostationary orbit, radio-frequencies used in space communication, solar energy, low earth orbits and Lagrange points, the internet, etc. The Arctic according to non-Arctic Asian countries is rightly called the 'common heritage of mankind' (Gautam 2011). However, it is an interesting thought process in an age of global governance and preventive policies.

6 Conclusion

The Arctic today is in an antithetical situation where, on the one hand, there are strong and important economic interests, and on the other, a need for climate protection and resource governance. The Arctic is viewed not as a wilderness anymore but as a region where the interplay between the geo-physical, geo-economics and geo-strategic is profound and the impact global. As the Arctic Ocean open up, new shipping routes are being explored and as the Arctic waters warms, the migration pattern of fish population is undergoing change. The melting of the sea ice and of the Greenland ice cap will not only influence the livelihood of the northern people but also influence the planetary climate system in several ways, among these being rising sea levels. Climate change is influencing the livelihood of northern peoples in both positive and negative ways. These physical changes and the new development opportunities have turned the Arctic into an increasingly important region in political terms.

The Asian countries are not untouched with the changes in the Arctic and are taking a strategic view of the Arctic affairs. Except for India, and to some extent Singapore, the other Asian countries' interests are tied strongly with the immediate economic developments in the Arctic region particularly through the opening of the NSR, which presents opportunities. In the years to come the Far East Asian countries involvement with the Arctic will only increase. But there is a downside which cannot be ignored. The melting of the arctic sea ice can result in adverse impact on the key global oceanic and atmospheric currents. While in the short run economic opportunities are arising, it is not clear whether in the long run the rapid melting of the Arctic ice is a positive development. Asian countries (The Arctic Council provides Observer status to non-Arctic states, to inter-governmental and inter-parliamentary organizations, as well as non-governmental organizations) should, therefore, coordinate in mitigating the likely adverse impact of the Arctic sea ice melt and also evolve a coordinated role in the Arctic Council through the six working groups that include Arctic Contaminants Action Program (ACAP); Arctic Monitoring and Assessment Programme (AMAP); Conservation of Arctic Fauna and Flora (CAFF); Emergency Prevention, Preparedness and Response (EPPR); Protection of the Arctic Marine Environment (PAME); Sustainable Development Working Group (SDWG), to strengthen the institution to the rapidly changing Arctic region since the Arctic is now an important geographical categorization and a critical part of the 'global knowledge commons'. As climate continues to change, and as actions are being forged, it is important to re-conceptualize and re-imagine regions from an ecosystem perspective. For example, South Asia should be seen as a 'riverine neighbourhood' or a 'Himalayan watershed' or a 'monsoon Asia' or a 'glacial Asia'. Such kind of conceptualisation is a departure from the conventional ways of dividing the world into political-geography to using specific ecosystem to understand regions as one organic continuum and as an organic cultural whole. Viewing seas as a great "commons" or considering Himalaya as the ultimate watershed or charting the monsoon as rain-dependent phenomena or studying the Arctic from snow, water, ice and permafrost perspective gives us a new lens to look at the world. It is said, "Man and not nature initiates, but nature in large measures control". Ideas will thus have to meet the dictates of ecosystem.

References

Gautam PKC 2011 The Arctic as a global commons. IDSA Issue Brief, 2 September 2011. Available at: http://www.idsa.in/system/files/IB_ArcticasaGlobalCommon.pdf

Gill A, Svigny D 2015 Sustainable Northern Development: the case for an Arctic Development Bank CIGI Papers, No. 54. Available at: https://www.cigionline.org/sites/default/files/no54.pdf

Hansen J, Ruedy R, Sato M, Lo K 2009 If it's that warm, how come it's so darned cold? Available at: http://www.columbia.edu/~jeh1/mailings/2010/20100115_Temperature2009.pdf

Hasnain SI 1967 United Nations environment program committee on global assessment of black carbon and troposphere ozone. Available at: http://www.unep.org/dewa/portals/67/pdf/BlackCarbon_report.pdf

IDSA 2013 The geopolitics of the Arctic: commerce, governance and policy. AsiArctic conference, September 23, the Institute for Defence Studies and Analyses (IDSA), New Delhi. Available at: http://www.idsa.in/event/TheGeopoliticsoftheArctic

IPCC 2013 Summary for policymakers. In: Climate change 2013: the physical science basis. Contribution of working group I to the Fifth assessment report of the intergovernmental panel on climate change. Cambridge University Press, Cambridge, United Kingdom and New York, NY, USA

Lü J, Hua J, Kim S, Ren J, Zhu Y 2008 Arctic oscillation and the autumn/winter snow depth over the Tibetan Plateau. J Geophys Res Atmos 113(D14):D14117

Østhagen A 2013 Arctic oil and gas: assessing the potential for hydrocarbon development in the polar region. Ottawa Life Magazine. Available at: http://www.ottawalife.com/2013/07/arctic-oil-and-gas-assessing-the-potential-for-hydrocarbon-development-in-the-polar-region/

Overseas Territory Review 2011 New initiatives could improve EU-Arctic Relations. Available at: http://overseasreview.blogspot.com/2011/04/new-initiatives-couldimprove-eu-arctic.html

Science Now. Is global warming making Tibet dustier? 2011 Available at: http://news.sciencemag.org/sciencenow/2011/01/is-global-warming-making-tibet-d.html

Sinha U (2013) The Arctic: challenges, prospects and opportunities for India, Debate. Indian Foreign Aff J 8(1):23

Saran S 2011 Why the Arctic Ocean is important to India. Business Standard, 12 June. Available at: http://business-standard.com/india/news/shyam-saran-whyarctic-ocean-is-important-to-india/438716/

UNITAR (United Nations Institute of Training and Research) 2013 Introductory e-Course to the global commons. Available at: http://www.unitar.org/event/introductory-e-courseonglobal-common

Should India Encourage Tourism in Antarctica?

Rasik Ravindra

Abstract The global tourism, in Antarctica, has been growing rapidly due to persistent curiosity and splendour of Antarctic region. Antarctica remains awe inspiring pristine land of unparalleled beauty that has to be seen and explored to be believed—for most of the people. Ever since the regulations of Environmental Protection measures became an essential component of tourism policy in Antarctica, the tourism industry has grouped itself under an organized non-governmental activity and arranged to adhere to CEP (Committee on Environmental Protection) guidelines on landings, visits to sites of historical and intrinsic values and maintaining safe distance from the wild life. This has encouraged State sponsorship from some countries in the form of issuing permits for visits to Antarctica which in turn has had visible contribution in the development of travel market and ancillary industries linked to shipping, port development etc., in the gateway centres that act as departure points for Antarctica. The tourist inflow to Antarctica is essentially concentrated to Antarctic Peninsula, South Shetland Islands and sub-Antarctic islands which are accessible from Argentina's port of Ushuaia or Punta Arenas of Chile by air. Cape Town, Christ Church and Hobart are other gateways to Antarctica. For tourists from northern hemisphere, especially Asia, Antarctic travel is expensive since visitors have to undertake a costly journey to South America to avail a cruise or air trip to Antarctica. It is opined that India should take a lead and fill up the geographical gap existing in this arena. There exists a lot of potential for India to support and encourage regulated tourism to Antarctica, especially as it has two operational research stations and one abandoned historical site of its first permanent station in Antarctica.

Keywords Environmental protection · Organized tourism · Gateways

---

R. Ravindra (✉)
Formerly with Ministry of Earth Sciences (MoES), New Delhi, India
e-mail: rasikravindra@gmail.com

P.S. Goel et al. (eds.), *Science and Geopolitics of The White World*,
DOI 10.1007/978-3-319-57765-4_5

1 Introduction

Exploration of Antarctica commenced with epic, second sea voyage of Captain Cook to Southern latitudes, during 1772–1775. Though he could not cross the great ice barrier surrounding the continent and so was not able to sight the "*great southern land mass or terra Australis incognita*" yet the discovery of several Sub-Antarctic Islands and rich marine life abounding there, encouraged several subsequent expeditions to hunt whales and seals and for establishing of stations on these islands. The landing on the continent itself was successful only in 1820.

The early years of twentieth century saw several expeditions to Antarctica, including the famous expeditions of Scott and Amundsen to South Pole, but most were exploratory adventures in nature, contributing to establishing geography of the continent. The major interest, world over was, however, developed in Antarctica essentially due to coordinated efforts of the countries participating in the International Geophysical Year in 1957–58, that also saw signing of Antarctic Treaty.

Tourism in Antarctica is comparatively a recent development that dates back to early 1950 (Headland 1994) with Argentina and Chile organizing short air and sea cruises to Peninsular parts of the continent. Since then, tourism has seen both increase in volume and diversity of tourist activity in Antarctica. The most common type of tourist activity continues to be ship based tourism that includes both landing and non-landing types. In recent years, yacht based tourism, adventure tourism (marathons, cycling, sky diving, mountaineering) and air based tourism have been added to the list. The latter includes over flights and air transport of group of tourists to and from some selected locations where ice run ways are available nearby. However, the ship based tourism still remains the most preferred mode. The International Association of Antarctic Tour Operators (IAATO) has been playing a significant role in bringing major tour operators under its umbrella so that a reliable control mechanism and data on such activities is being made available to Antarctic Treaty secretariat.

2 Growth of Commercial Aspects in Antarctica

Antarctica was too far, too risky and inaccessible for any commercial venture apart from the exploitation of whales and seals in late nineteen and early twentieth century. The over exploitation and unregulated killings soon made the whaling industry uneconomical and resulted in abandoning of the whaling stations in sub-Antarctic islands. The discussions about possible resource exploitation in Antarctic surfaced in 1970 and initiation of negotiation for "convention on the regulations of the Antarctic Mineral resource Activities" (CRAMRA) along with media coverage hinting that Antarctica might have vast mineral resources fuelled the economic interest once again in this land. The CRAMRA was finally concluded on 2nd June 1988 and put a halt to aspirations of mining companies. The only activity that could be cashed in for gains, left was thus tourism.

The tourist inflow to Antarctic was practically negligible in Sixties and started picking up in Seventies when maximum tourists preferred flying over the accessible parts of Antarctic Peninsula. The commercial ship based tourism commenced in 1965–66 year with the Cruise ship *Lindblad Explorer* taking passengers to Antarctic Peninsula. The shortage of ice breakers was a great impediment in growth of tourism in Antarctica but 1990 onwards several ice breakers, from erstwhile USSR, became available to give a boost to the tourist activity. This saw an exponential growth in tourism between 1990 and 2010 from about 5000 tourists a year to more than 35,000 tourists in 2007–08. However, the economical recession during 2010 and 2012 saw numbers of visitors to Antarctica dipping to around 20,000 (Fig. 1). During the same time revised regulations of IMO, about ban of

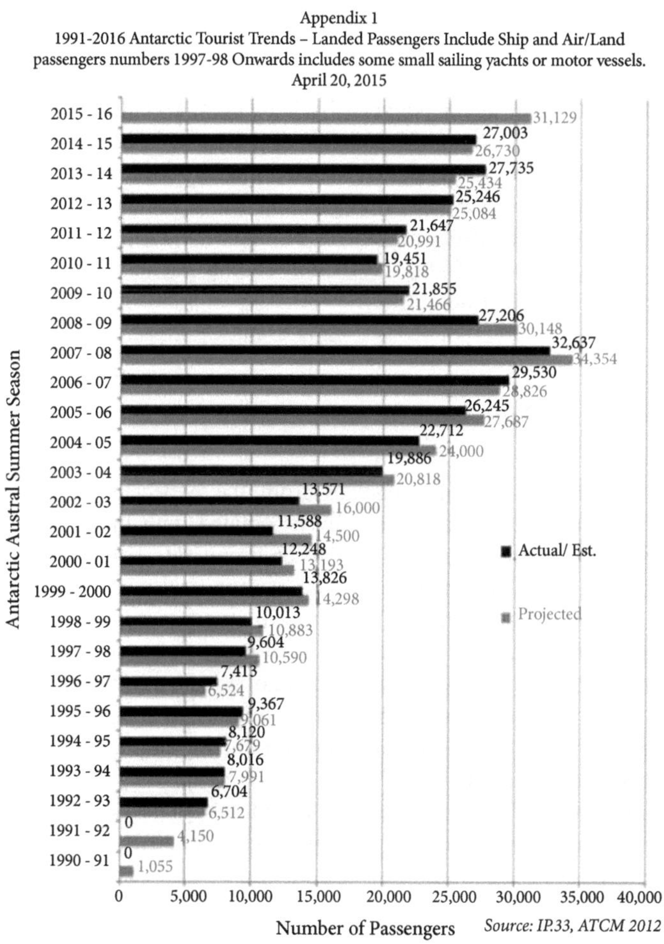

Fig. 1 Trends of tourists visits to Antarctica

Table 1 Diversity of activities and number of tourists

Type of activity	Number of vessels	Number of departures	Number of tourists
Tourist to peninsula (Seaborne)	22	193	34,329
Tourists to Ross Sea Region	5	7	499
Tourists in sailing boats/yachts	20	49	368
Tourists in air cruises to Peninsula	3	27	1848
Deep field tourism	–	–	361

Source IAATO Information Paper 22, ATCM, 2012

heavy diesel oil in the Antarctic Treaty Area (south of 60° latitude), also forced some small cruise operators to withdraw due to financial burden that the change from cheaper heavy oil. An unusual six numbers of the ship accident events between 2007 and 2009, in Antarctic waters may also have contributed to this low turnout. The visits have again bounced back in more recent years and projected number in 2015–16 was around 30,000 (IAATO Report 2012).

An analysis of actual tourist activity undertaken by Tour Operators registered with IAATO during season 2013–14, as given in Table 1, clearly demonstrates that seaborne tourism for Peninsula region far exceeds (34,329) other types, such as voyages to Ross sea sector (499), tourists in sailing yachts (368), Air cruises to Peninsula (1848) and deep field (361).

3 Concerns on Tourism

The main concern on tourism emanates from the expected adverse impact of anthropogenic activities that are generally associated with major involvement of human movement or presence in a pristine environment. The CEP and the environmentalist point out that Antarctica does not support the infrastructure which is required to restore an environmentally damaged site to its original form or to cater for pressure on land created by repeated visits of groups of visitors to a few selected sites. The question of disturbance to the penguins and other wild fauna such as petrels and skua that frequent such sites for nesting, breeding and hatching of eggs, also weighed high. The period of mating and colonizing of penguins also coincides with the peak tourist season.

Since some areas of Antarctic Peninsula and sub-Antarctic islands experience milder climate than the main land of Antarctica, the birds as well as the Tour Operators prefer such locations, resulting in high "landfalls" on these locations those results in increased foot prints (even walking on moss beds would leave

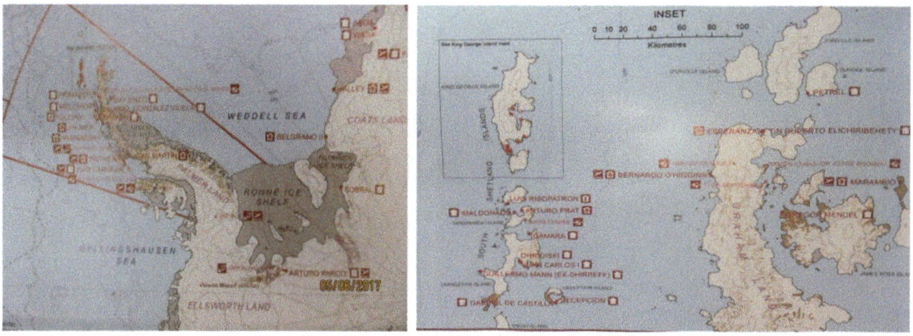

Fig. 2 Antarctic Peninsula and sub-Antarctic islands frequented by tourists. *Source* COMNAP

footprints that would last for decades, if not centuries), apart from causing some damage to heritage/historical sites.

Some of the more frequently visited locations in Antarctica Peninsula namely; Turtle Point, Penguin Island, Half Moon Island, Whalers Bay, Brown Bluff, Devils Island, Jougla Point, Horseshoe Island, Baily Head, Telefon Bay are scattered from northern tip of Peninsula to about 680 South Latitude as shown in Fig. 2.

Environmentalists are also concerned about the introduction of non-native species in the Antarctic environment by the visitors knowingly or unknowingly. Soil, seeds, fungi and insects can attach themselves to materials being brought by visitors to Antarctica and prove to be the carriers of non-native species. The invasive types of these non-native species have potential of putting to risk the survival of native species. The introduction of an alien grass on King George Island and several new species on sub-Antarctic islands has been cited as example.

The newer means of attracting tourists by tour operators such as arranging events of adventure tourism has also been on rise. These events include climbing, skiing/snowboarding, ice walking, snorkelling and scuba diving, marathons, tracking, kayaking and can cause more damage to the fragile environment, as these involve camping on the land or ice and associated logistics.

The ship accidents (Fig. 3) too have been a focus of attention by those who oppose increased tourism in Antarctic waters. Apart from the possible human causalities, such accidents cause a possibility of oil seepage and disruption of scheduled activities of scientific Stations, who have to divert the Research Ships to rescue operations. Some of the major accidents that have occurred in recent past are:

- Nov, 2006, Quark expedition ship that sank near Deception Island
- Jan. 2007, Hurtgruten Nordkapp ship. It had 294 passengers who were rescued by a sister Norwegian ship
- 2007, A Canadian cruise ship sank near South Shetland Island. 154 passengers had to be rescued
- 2007, The Explorer sinks near KG Island. 100 passenger rescued
- Nov. 2007, the Farm. It lost power and drifted into iceberg

Fig. 3 The explorer sinking near King George Island and MV Academik Shokalskiy stuck in ice. *Source* Public Domain (www.google.de)

- Dec. 2008, the Argentine ship Ushuaia ran aground with 89 passengers
- Feb. 2009, The Ocean Nova grounded.

In addition to above, during end of December 2014, MV Academik Shokalskiy, a Russian expedition ship with Australian expedition members got trapped in pack ice and members had to be evacuated by a Chinese helicopter.

4 Why then, We Should Support Tourism

A natural question that would arise then is that if there are so many negative impacts on the environment, why should tourism be supported in Antarctica? The supporters of tourist activity in Antarctica furnish several arguments that are being summed up below:

- Tourism has got a vast educational value, especially for the younger generation. The exceptional environmental values preserved in Antarctica can be a study ground for public and students both, who then can play an increasing role in advocating and upholding the intrinsic values of Antarctica.
- Without a native population of its own, Antarctica needs advocates to highlight need of a safe tourism. Tourism creates a global community of people that would go all the way to support—and indeed fund tourism and may be preservation of some heritage sites that are historical in nature. Unless people visit the Continent, they will not have firsthand experience to talk on the issue.
- "It is better to have a certain level of responsible tourism". In fact, research suggests that scientific programs may have much more environmental impact than tourism, because of its permanent presence.
- Though during 2015, about 37,000 tourists are expected to travel to Antarctica, but more than 10,000 among them will never go ashore.

- Ever since the regulations of Environmental Protection measures became an essential component of tourism policy in Antarctica and guidelines/check lists were issued, the tourism industry has grouped itself under an organized non-governmental activity and arranged to adhere to CEP guidelines on landings, visits to sites of historical and intrinsic values and maintaining safe distance from wild life.
- About half the tourist ships that go to Antarctica are flagged by Treaty nations. Most of the tour operators bringing tourists to Antarctica from these countries are members of IAATO, which works closely with the Antarctic Treaty System and are committed to follow all the guidelines prescribed by CEP. The other half are unfortunately private, on which Treaty has no control.

5 Indian Stand on Tourism in Antarctica (Suggested)

Recognizing the ground reality that tourism has come to stay in Antarctica, India advocated for a regulated tourism in Antarctica during the 2007 ATCM, held in New Delhi. However, there is no official policy on supporting or not supporting tourism apart from the above stand. India did, however, present an Information paper on Tourism in recently concluded ATCM that reflects India's keen interest and involvement in the issue.

The data provided by IAATO to ATCM show that the 36,271 tourists that visited Antarctica during 2014–15, belonged to as many as a hundred nationalities. Most of these visitors came from eight countries namely USA, UK, Australia, China, Canada, France, Germany, Switzerland and others with little contribution from India and Middle East countries. Lack of infrastructure facilities and governmental support may be one reason for this poor response. It is generally agreed that Antarctic tourism has got both educational and commercial advantage to mankind and therefore needs to be promoted, provided it is regulated, so as to ensure safety of environment and associated ecosystem of Antarctica.

It is opined that India should take a lead and fill up the geographical gap existing in this arena. There exists a lot of potential for India to support and encourage regulated tourism to Antarctica, especially as it has two operational research stations and one abandoned historical site of its first permanent station, Dakshin Gangotri. It has two Historical sites and manages two areas of ASPA and ASMA (Antarctic Specially Protected Area and Antarctica Specially Managed Area). There will be a natural curiosity and interest among Indians to look and capture the historical and current status of these sites apart from having a glance of the picturesque Antarctic scenes, ice bergs, penguins and other sea life.

Goa, which is an all time favourite destination for tourists and is the seat of a leading Indian R&D Institute for Antarctic research, can be developed as an Asian "gateway to Antarctica" to serve as a departure point for cruises. It has high potential for attracting local and foreign travellers to icy continent because of existing port and logistic facilities and scientific and technical know how required for such an activity. This would not only come as a boon to tourism industry in India but will also make travel financially viable for interested people as there will no longer be a need to spend on costly travel to South American gate ways.

As the Indian activities for Antarctica (expeditions) are managed from Goa, the Indian tourism industry can gain immensely from the expertise available with India due to its presence in Antarctica for more than three decades. The industry can get the local experts (guides) in the fields of science and logistics who also have geographical and historical knowledge of sites, wild life and biological diversity, since it is mandatory that all landings of tourists in Antarctica are made under supervision of experts.

References

Headland RK (1994) Historical development of Antarctic Tourism. Ann Tourism Res 21(2):269–280

IAATO Report (2012) Environmental aspects and impacts of tourism and non-governmental activities in Antarctica. IP 33, Antarctic Treaty Consultative Meeting XXXV, Hobart, pp 1–39

Part II
The Global Climate Change and Polar Region

The Arctic Teleconnections

M.N. Rajeevan

Abstract The Arctic Region is warming at an alarming rate due to snow/ice-temperature feedback as a result of increase in greenhouse gases and other anthropogenic activities. Due to the Arctic amplification of warming, the Arctic sea ice here is melting at a faster rate of about 0.67 million km^2 per decade. Past observational data suggest that the Indian Monsoon variability is linked to the Arctic Sea ice variability. In fact, about 10% of Indian monsoon variability is explained by Arctic Sea Ice. Further, the effect of sea ice decline has implications in changes of large scale circulation of mid-latitudes, for example a southward shift of Sub-tropical westerly Jet Stream, and occurrence of extreme weather events over Europe and Asia. The abnormal heavy rainfall spell during March 2015 over the northern parts of India could be linked to the Arctic sea ice melting and associated changes in mid-latitude circulation. The analyses of Arctic Circulation anomalies have revealed that the Arctic Oscillation (opposing pattern of pressure between the Arctic and the northern middle latitudes) and the Indian Monsoon are inter-linked. An interesting twist in the relationship between the Arctic and Indian monsoon, has been observed as a result of the recent studies, which shows that abnormal convection over the northwest India could influence Arctic sea ice melting through mid-latitude circulation anomalies and wave-guide.

Keywords Arctic warming · Arctic oscillation · Tele-connections

1 Introduction

The Arctic is defined as the region above the Arctic Circle, an imaginary line that circles the globe at 66°32″ N (Fig. 1). The Arctic influences the weather and climate of the entire Northern Hemisphere, and the cool northern region helps to moderate the climate of the rest of the planet. Arctic has been warming rather rapidly.

M.N. Rajeevan (✉)
Ministry of Earth Sciences, Government of India, New Delhi, India
e-mail: secretary@moes.gov.in

P.S. Goel et al. (eds.), *Science and Geopolitics of The White World*,
DOI 10.1007/978-3-319-57765-4_6

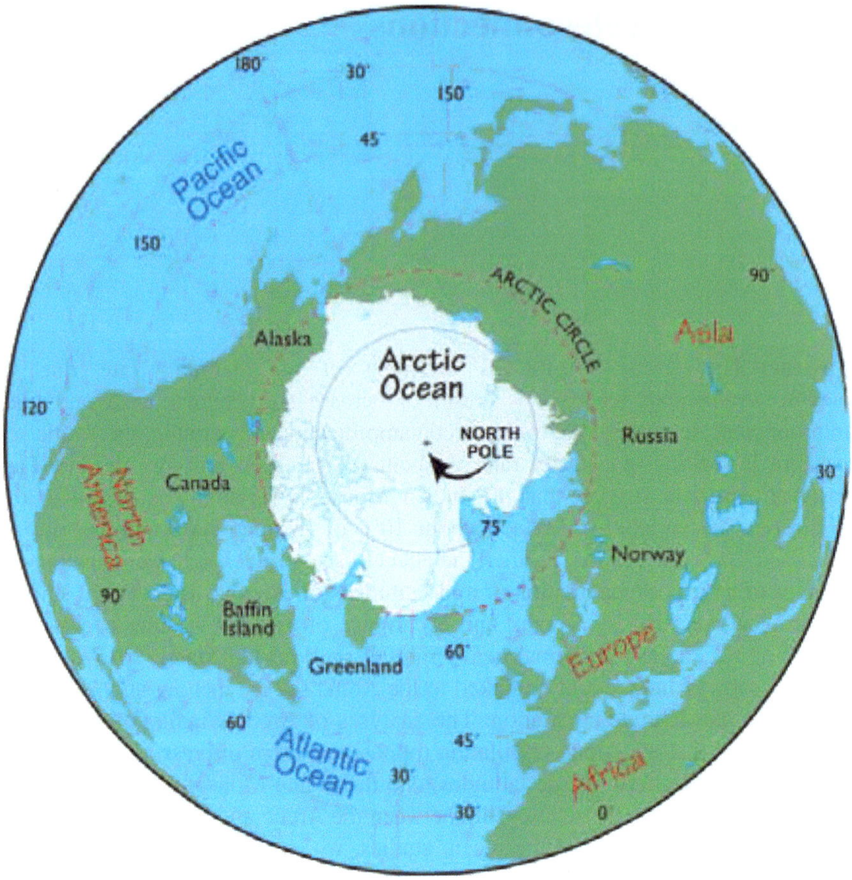

Fig. 1 The Arctic region. *Source* www.google.de

The Earth has witnessed warming since last 100 years, globe by about 0.6 °C, but in Arctic, the rate is much higher i.e., 2 °C warmer in the last 100 years. The difference is large, mainly due to the decline of the sea ice that covers the Arctic sea, while other factors such as changes in cloud cover, increases in atmospheric water vapour, and atmospheric heat transport from lower latitudes may also be contributing factors. Screen and Simmonds (2010) have emphasized that the diminishing sea ice in recent years has been the main cause of the Arctic amplification.

2 Diminishing Sea Ice Over Arctic

The recent IPCC report has confirmed that decrease in the sea ice trends as reported in the AR 4 has continued. The annual Arctic sea ice extent has decreased over the period 1979–2012. The rate of this decrease has been computed at between 3.5 and

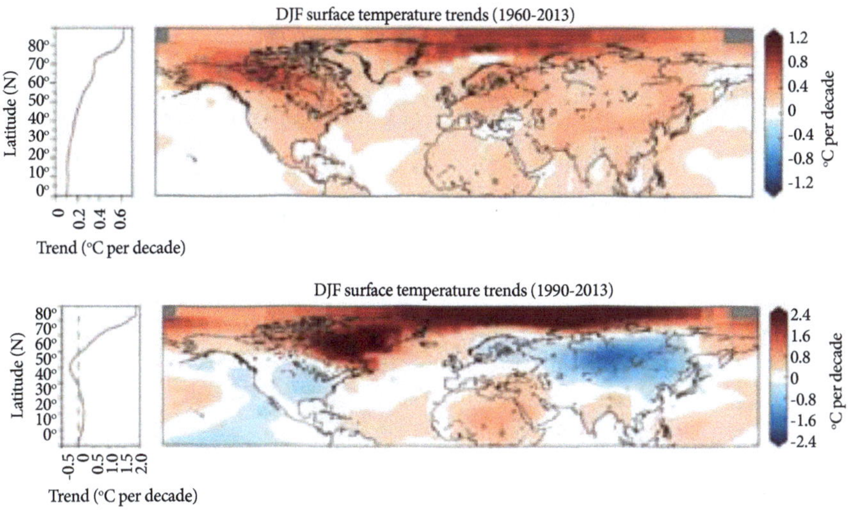

Fig. 2 The global temperature and annual temperature trends. *Source* Cohen et al. 2014, Nature

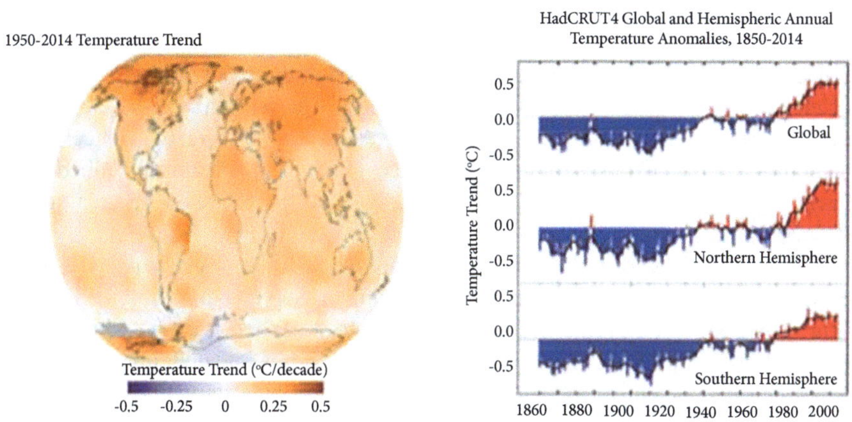

Fig. 3 The Arctic amplification. *Source* P. D. Jones, T. J. Osborn and K. R. Briffa, University of East Anglia Norwich, UK, D. R. Parker, Met Office, Bracknell, Berkshire, UK

4.1 per decade, with the average decrease in decadal extent being most rapid in summer and autumn though there has been a consistent decrease in every season. The multiyear sea ice too has decreased between 1979 and 2012. This is supported by the satellite observations.

The trend of the exceptional warming in last 54 years in Arctic as compared to the other parts of the world is seen in Fig. 2. The warming is much more over the Arctic region as compared to the global average. A closer examination of Fig. 3, one can see that the trend for two different places, one 1960–2013 and 1990–2013.

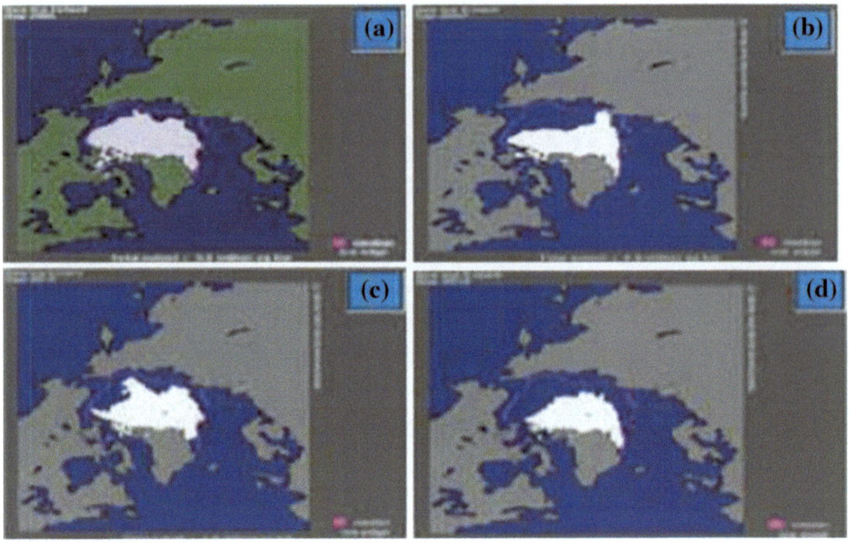

Fig. 4 Arctic Sea ice decline. Clockwise, the sea ice extent in September 2005, September 2007, September 2012 and September 2010. *Source* NSID Centre Data

It is evident that the warmer trends are pronounced in the more recent years and the change in rate is almost 2 °C in the last 20 years. The bottom line therefore being that the rate in increase of the warming of Arctic region is a big concern. The depleting sea ice exposes more sea surface to the sun, decreasing albedo and leading to absorption of heat by the sea water, thereby increasing the sea surface temperature.

In the ideal cycle of the Arctic sea ice—the annual cycle—the maximum extent is seen in February–March and the minimum in September. The satellite data (Fig. 4) shows that extreme decline in the extent of sea ice over the Arctic region occurred in 2005, 2007, 2010 and 2012. Figure 5 exhibits updated time series of the Arctic sea ice content from 1979, when the satellite's data started coming, and till 2014. In the initial few years the trend is not very high but 1997 onwards the trend is rather fast. From 1979 to 2013, the sea ice extent has decreased by 4 million square kilometres which is a very large change in the sea ice content over the Arctic region.

Arctic oscillation, yet another atmospheric phenomenon (just like ENSO) also is important. Though it is not periodic, *senso strictly*, but is still referred as periodic oscillation between the polar region and the mid-latitude region. It has large implications in the regional as well as the global climate. The Arctic oscillation has two phases (Fig. 6). One is the positive phase in which the polar vortex, the cold stratosphere are very strong, can cause strong storms moving over northern parts of Europe. In the second, which is the negative phase, we get less cold stratosphere and less amplitude of the storms. The melting of sea ice, basically leads to more

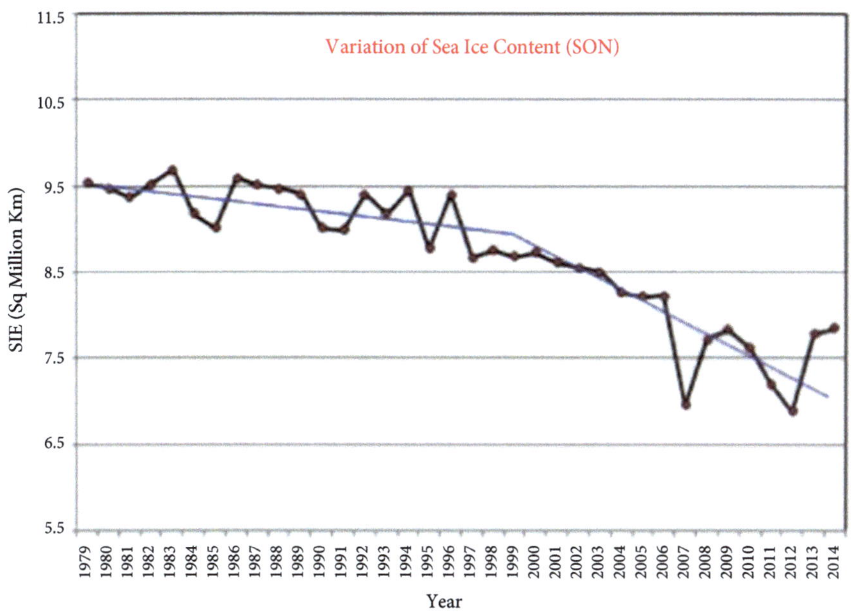

Fig. 5 Sea Ice extent 1979 onwards. *Source* NSID Centre Data

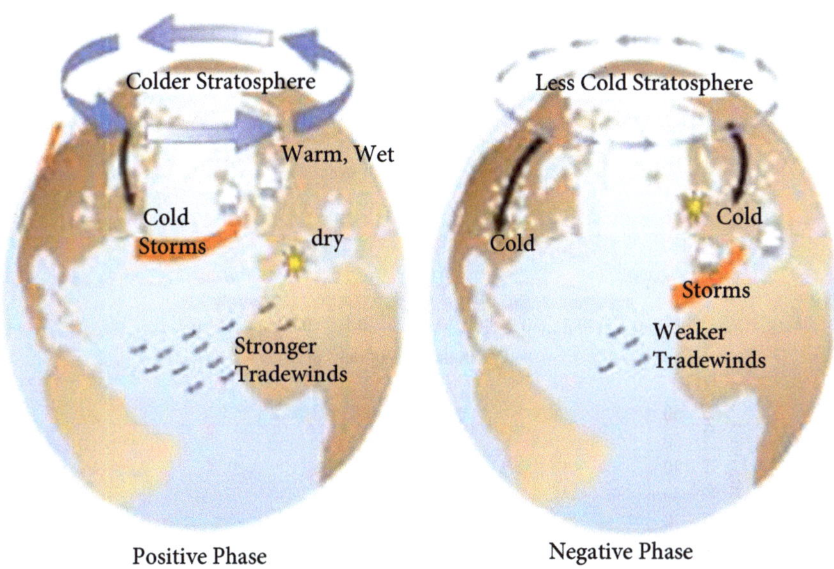

Fig. 6 The Arctic oscillation. *Source* www.earthweek.com

frequent negative phases of the Arctic oscillation. Melting of sea ice causes imbalance in the radiative budget in the region.

Surface Albedo change causes radiative budget. The impact will pronounce in the atmosphere and its circulation. The possible changes related to Arctic amplification and teleconnections can be in terms of changes in storm tracks, jet stream and planetary waves. In the mid latitude there is a large scale planetary waves which is moving from west to east in the higher atmosphere level. The analyses of last few years data, has shown that the jet stream in the northern hemisphere comes down in

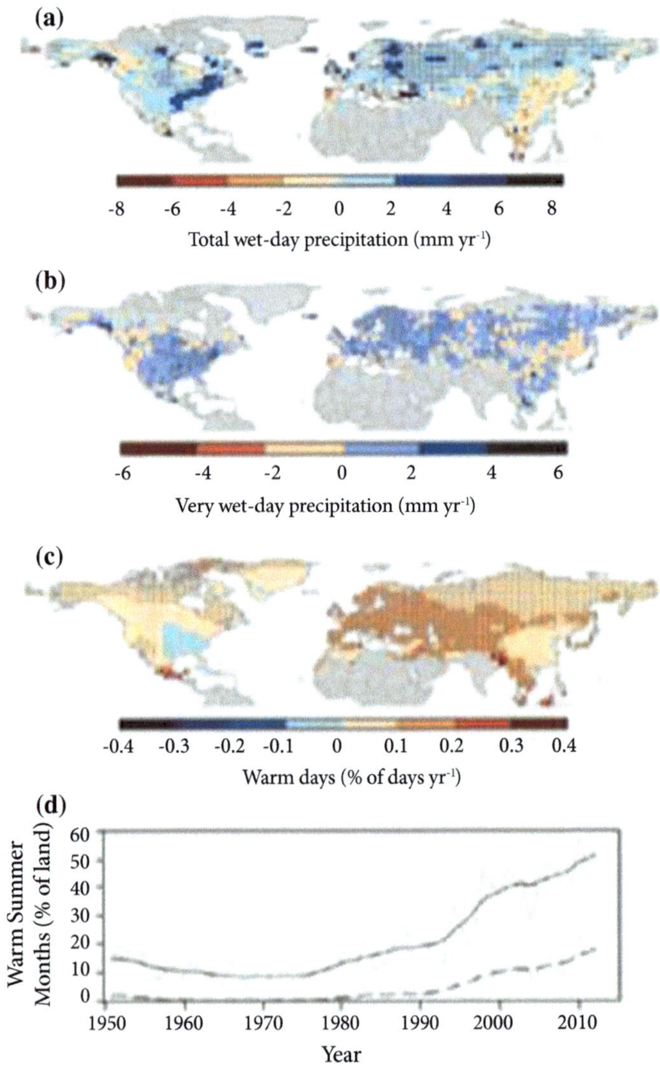

Fig. 7 Trends during Trends during 1951–2013. *Source* Cohen et al. 2014, Nature

the winter and pre-monsoon season, which has large implications over the European region and southern part of Asia.

One can see the trends in Europe's extreme weather events as depicted in Fig. 7. The upper top shows precipitation and the bottom part shows the warm events. Some of these may be the result of Arctic sea ice melting as well as Arctic amplification.

3 Arctic—Monsoon Teleconnections

Some analysis of the relationship between the Arctic sea ice melt and Indian monsoon has been attempted by Indian scientists. There is a good correlation between arctic sea ice and Indian monsoon. About 8–10% of monsoon variability is explained by this relationship. We have also associations with the Arctic oscillation and Indian monsoon which is not as strong as the sea ice extent relationship.

The negative phase of Arctic oscillation which is shown in the sea level pressure as well as temperature also increases the snow cover in the north-western parts of the Europe which also has an implication over the India monsoon. The Arctic teleconnection enhances frequency of blocking highs in the mid latitudes, which has large implications. For example, the Pakistan floods of 2010, where thousands of people had died, were triggered by the blocking highs in the atmosphere which was seen at 5–6 km level. At the same time, Russia had a huge heat wave which lasted for about 15 days, both the events were cast by the same weather systems. For Russia it was the heat wave, whereas in Pakistan it was a flood.

During March, 2015 there was a huge rainfall activity and hailstorms which also happened in the immediate preceding year i.e. in 2014. All blue lines in the Fig. 8 are the areas where the rainfall was in excess during March, which resulted in the

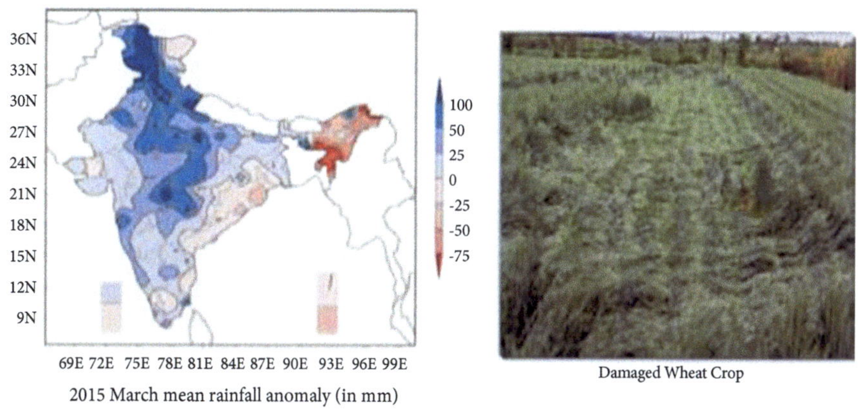

Fig. 8 Arctic connections for Abnormal Wet March 2015? *Source* Climate Research and Services, IMD, Pune

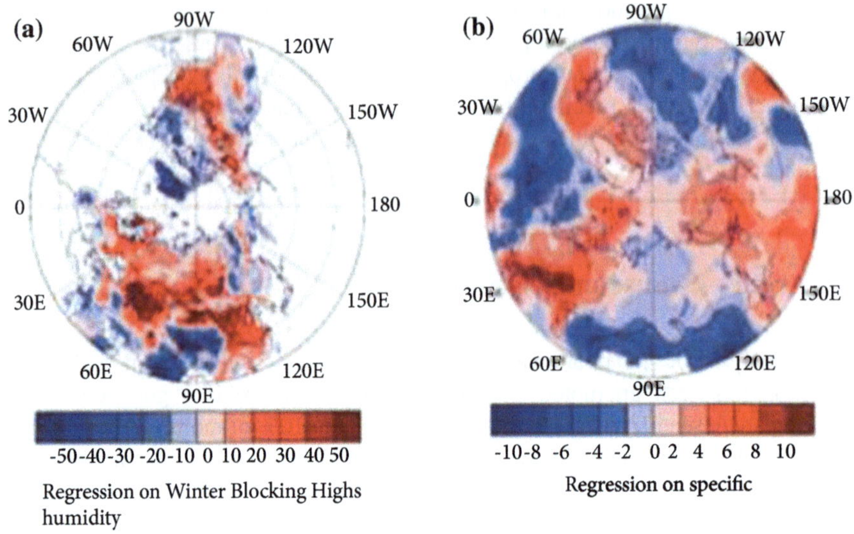

Fig. 9 Arctic teleconnections enhance frequency of blocking highs over Eurasia. *Source* Liu et al. PNAS, 2012

damaged Wheat crop, due to untimely rains during this pre-monsoon season. Our analysis clearly shows that this could be linked to the teleconnections with Arctic sea ice melting. The whole jet stream is pushed down and associated with the jet stream, this weather event had happened. In the north of the jet stream, there was a blocking high. This kind of frequent atmospheric setting is taking place in the last 3–4 years which could be related to sea ice melting.

In the right hand side of Fig. 9, a blocking high over the northern parts of Europe is seen. This could be the causes of pre-monsoon rains in India. Left top of the slide, one can see a blocking high and over India, it is a trough. Basically, it is a large planetary wave locked to the region and this caused the heat wave in Russia and the same Weather system affected Pakistan with floods. Recent studies suggest that it could also be a signal from the Arctic teleconnections. We know that in 2015 May and June, heat waves occurred over India especially in Andhra Pradesh in which many people died.

On the left hand side of the Fig. 10, heat waves over central and northern parts of India are depicted while on the right hand side of this figure the maximum duration of the event is shown. One can see the substantial increase in the maximum durations as well as the average durations of heat waves over India. Our preliminary analysis says that it also got a signal form Arctic sea ice and amplifications.

If we analyse the Arctic oscillations with the extreme heat waves, we see a relationship between heat wave frequencies and Arctic Oscillation over India.

Dr. T.N. Krishnamurti, one of the leading scientists in US, working in Florida state university; has made an analysis and suggested that these kind of events

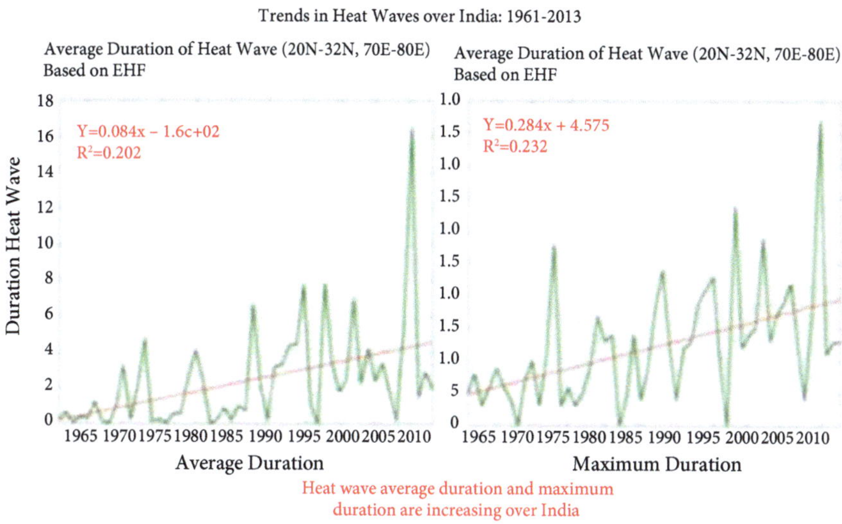

Fig. 10 Increasing trend of Average and maximum duration of Heat wave. *Source* Author (Sc paper: Rohini, Rajeevan and A. K. Srivastava)

(e.g., the Pakistan floods of 2010), can really trigger more melting in the Arctic sea ice. Heavy rainfall in India can really increase the latent heat release and that energy can propagate to the Arctic region through planetary waves.

The hypothesis he is making is that the heavy rainfall over north-west India and the Himalayan region can release a lot of latent heat over the atmosphere which can be propagated to the Arctic region. It can also enhance the natural melting of the sea ice. These are some of the grey areas where further research is required.

In conclusion, the studies do indicate a relationship between the teleconnections of Arctic climate variability as well as the regional climate variability.

References

Cohen J, Screen JA, Furtado JC, Barlow M, Whittleston D, Coumou D, Francis J, Dethloff K, Entekhabi D, Overland J, Jones J (2014) Recent Arctic amplification and extreme mid-latitude weather. Nature Geoscience

Jiping Liu, Curry Judith A, Wang Huijun, Song Mirong, Horton Radley M (2012) Impact of declining Arctic sea ice on winter snowfall. Proc Nat Acad Sc (PNAS) 109(11):4074–4079. doi:10.1073/pnas.1114910109

Screen JA, Simmonds I (2010) The central role of diminishing sea ice in recent Arctic temperature amplification. Nature 464:1334–1337. doi:10.1038/nature09051

Abrupt Climate Shifts Over the Past 10,000+ Years: An Arctic-Antarctic-Asian Imbroglio?

S. Rajan

Abstract The Asian Monsoon climate system and accompanying precipitation play a significant role in large-scale climate variability and water supply over much of the globe. For instance, the Indian agriculture, which accounts for ~25% of the GDP and employs over 70% of the population, is heavily dependent on the monsoon rains, 80% of which arrives during the summer. While our predictive capabilities of the monsoon have improved by leaps and bounds over the past few years, perhaps a major missing link in our understanding of the phenomenon of the monsoons is knowledge of the forcing functions behind the short-term monsoon variability. To predict the evolution of inter-regional climate linkages on decadal and shorter time scales, it is crucial to understand how they evolved in the past. Palaeoclimatic studies using proxies from the marine and terrestrial records present large variations of the monsoon systems during the last glacial period and over the Holocene, which can be linked to abrupt millennial-scale warm-cold episodes over Greenland and the North Atlantic. For example, speleothem records from China indicate the links between Asian monsoon and solar input and North Atlantic ice-rafting events, during the Holocene. The Chinese records also appear to correlate fairly well with corresponding short-term variations in the Indian summer monsoon. Studies of the annual sea-ice variability over Antarctica also seem to indicate that deficient monsoon years are preceded by more than normal sea-ice extent and vice versa. Has such a linkage between the Polar Regions and the Asian monsoon persisted through time and if so, what has (have) been the forcing function (s)? Is there a synchronous or time-lagged inter-hemispheric linkage between the Arctic-Antarctic and the Asian monsoon? In this paper, I focus on these and related questions and highlight the importance of understanding the role of polar and high latitudinal regions on the monsoon phenomena.

Keywords Monsoon variability · Climatic linkages · Palaeoclimate

---

S. Rajan (✉)
National Centre for Antarctic and Ocean Research, Vasco da Gama, Goa, India
e-mail: rajan.ncaor@gmail.com

P.S. Goel et al. (eds.), *Science and Geopolitics of The White World*,
DOI 10.1007/978-3-319-57765-4_7

1 Introduction

Proxy records of the climate system spanning the last 100,000 years and beyond demonstrate abrupt climatic switches at millennial-scale (10^3–10^4 years) frequencies, which are superimposed upon the orbitally-modulated 20–100-k year Milankovitch band cycles of insolation (cf. Dansgaard et al. 1993). Changes of up to 16 °C and a factor of 2 in precipitation have occurred in some places in periods as short as decades to years (e.g. Broecker 1994, 1995). Perhaps the best-known and probably the best-dated abrupt climate change event of the last 100,000 years is the Younger Dryas (YD), a brief ($\sim 10^3$ years), cool (cold?) climate episode straddling the transition from the last glacial period to the relatively gradual warming trend of the Holocene (roughly the last 10,000 years). Although a relatively stable regime as compared to the last glacial period, the Holocene was also punctuated by many abrupt climate change events of a lesser amplitude (Overpeck 1996). The forcing functions behind such abrupt climate changes superposed on a cooling phase of the last glacial period or the gradual warming trend of the Holocene are, however, matters of academic debate. As discussed later, many hypotheses do exist, and there is strong evidence of change in the fundamental mode of operation of the coupled land-ocean-atmosphere-cryosphere system. Understanding the causes of both types of abrupt climate change is essential for assessing the importance of their role in our climate future.

The dominantly agrarian economy of Asia and in particular, of the Indian region is overwhelmingly dependent on the seasonal reversal of both the monsoon surface winds and attendant precipitation. For instance, the Indian agriculture, which accounts for $\sim$25% of the GDP and employs over 70% of the population, is heavily dependent on the monsoon rains, 80% of which arrives during the summer. The Asian monsoon system comprising the Southwest Asian or the Indian summer monsoon (ISM) and the East Asian summer monsoon (EASM), also plays a vital role in large-scale climate variability of the region in terms of transportation of heat and moisture from the tropical ocean across the equator to higher latitudes (Wang et al. 2001). Available evidence also seems to suggest that on century-to millennium time scales, the strength of the Asian monsoon has varied significantly (Schulz et al. 1998; Sirocko et al. 1999; Wang et al. 2001; Altabet et al. 2002) and that these changes are, for some reason, correlatable with the millennial-scale climate events in the North Atlantic region. The Indian monsoon has also been suggested to vary on these timescales (Overpeck et al. 1996; Schulz et al. 1998; Kudrass et al. 2001; Gupta et al. 2003).

From the foregoing it can be seen that while the earth's climate history spanning the last 100,000 years and beyond is marked by centennial-to millennial scale abrupt climate changes, perhaps a major missing link in our understanding of this phenomenon is a knowledge of the forcing functions behind these abrupt changes as well as their spatial and temporal characteristics. To predict the evolution of the inter-regional climate linkages on shorter time scales, it is crucial to understand how they evolved in the past. This is especially significant as regards our predictive capabilities of the monsoon, considering that the regional monsoon system affects more than half of humanity worldwide. Credibility of future monsoon projections is however, limited by the large spread in the simulated magnitude of precipitation changes, poor agreement between models, and the short and sparse Instrumental records. The uncertainties in predictions can be minimized to an extent by focusing on the constraints provided by proxy records of long-term climate trends and shorter-term variability. The past 10,000 odd years of climate history is ideal in this context because this interval is characterized by changes in the distribution of incoming solar radiation, giving rise to changes in both the long-term mean monsoon and in short-term climate variability.

The present study summarizes our current understanding of the abrupt climate changes over the past 10,000+ years, focusing on the proxy records of monsoon variability, and their probable linkages with correlative climate change events of the Polar Regions. Some of the key questions that arise in this context and which call for the attention of the palaeoclimate community are:

- Do the Polar Regions drive/modulate/have a say in the global climate and monsoon? (Did they in the past or will they in the future?)
- If so, what have been or what are the forcing functions?
- What is the likelihood of abrupt or critical climate and/or earth system changes resulting from processes in the Polar cryosphere?
- What will be the impact of changes in the Polar cryosphere on the atmospheric and oceanic circulation?
- What will be the nature of changes in sea-ice distribution and mass balance in response to climate change and variability?

2 Asian Monsoon Variability During the Holocene

Palaeoclimate records spanning the last glaciation point to an overall drier climate and weaker monsoon systems over much of Asia, probably on account of the presence of large ice sheets and reduced atmospheric CO_2 concentration. Modelling experiments also show that a reduction in moisture transport resulted in less precipitation over East Asia during the Last Glacial Maximum (Yanase and Abe-Ouchi 2007).

During the subsequent Holocene period of gradual warming, the strength of the Asian summer monsoon followed the insolation patterns, with increased summer monsoon intensity during the early part, and a gradual decline from the mid-Holocene onwards, corresponding to the long-term decline in summer insolation (Kutzbach 1981; Wang 2005) (Fig. 1). This trend is however, not synchronous across the different monsoon regimes. The timing of the strengthening and weakening of the Asian monsoon varies significantly among different sites as shown by the palaeo-records across Asia. The proxy records also show abrupt decadal to centennial variations superimposed on this long-term weakening trend of

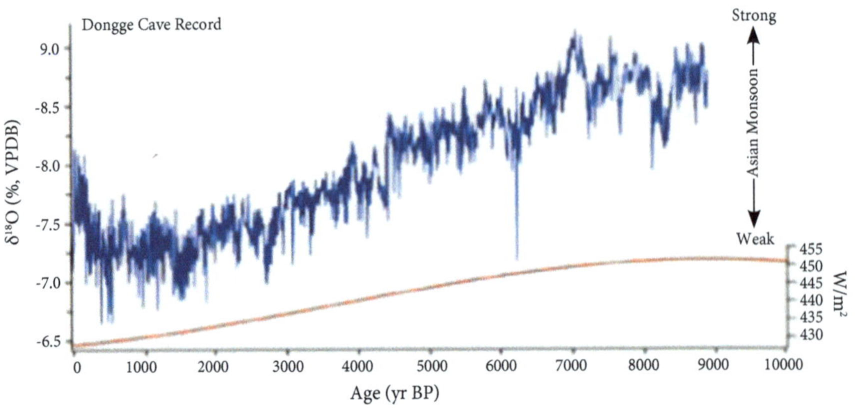

Fig. 1 Proxy record from the Dongge Cave stalagmite, China. *Source* Adapted from Chawchai (2014), Wang (2005) and Laskar et al. (2004)

Holocene. Various hypotheses have been proposed to explain this decadal to centennial scale variability, such as solar forcing, the El Niño Southern Oscillation (ENSO) and Indo-Pacific climate variability, movement of the mean position of the ITCZ, and changes in the Indian Ocean Dipole etc., (see, Chawchai 2014).

3 Indian Summer Monsoon Variability During the Holocene

Fairly complete records of abrupt low-latitude climate change events during the last 100,000 years come from the marine proxy records of northern Indian Ocean in the Arabian Sea and the Bay of Bengal, and from the stalagmite records (Schulz et al. 1998; Kudrass et al. 2001; Burns et al. 2003; Marzin et al. 2013). The pattern of Indian Summer monsoon precipitation in these records bears a striking resemblance to the oxygen-isotope record from Greenland ice cores, with increased precipitation associated with warm periods in the northern latitudes (Kudrass et al. 2001).

Despite certain temporal and spatial variations, the proxy records of Indian Summer Monsoon during the Holocene largely mimic the general trend of corresponding variabilities in the Asian monsoon records. For instance, sediment records from the Arabian Sea demonstrate a pattern of persistent variability in the Indian Summer monsoon throughout the Holocene that may be linked with episodic warming/cooling of the North Atlantic (Gupta et al. 2003). Speleothem records from the Dandak cave of India demonstrate the presence of high-amplitude, recurrent multi-decadal scale pattern of climatic variabilities between the 7th and 16th Centuries A.D. This record displays a high degree of correlation with a speleothem-based Asian monsoon reconstruction from Wanxiang Cave of China testifying to the regional significance of these short-term climate variability (Berkelhammer et al. 2010).

4 The Link Between Asian Monsoon and the Polar Regions

Continuous profiles of the oxygen isotope ($\delta^{18}O$) measurements and variations in the dust content from Greenland ice cores point to the general climate instability of the Northern Hemisphere over the past 250 k years (Dansgaard et al. 1993; GRIP Project Members 1993). For instance, the ice cores corresponding to the last glacial period exhibit a string of more than 24 large, abrupt climate perturbations when the temperatures increased by up to 15 °C relative to the full glacial values. These short-term ice core climate oscillations, which have come to be known as "Dansgaard-Oeschger (D-O) events or D-O Oscillations" have also been identified in high-resolution climate records from the Antarctic ice cores, and in the sediments from the North and South Atlantic, northeastern Pacific and northern Arabian Sea etc. (Keigwin et al. 1994; Behl and Kennett 1996; Charles et al. 1996; Schultz et al. 1998), testifying to their global imprints. Sediment cores from the North Atlantic show that the D/O events are bundled into larger (sea surface) cooling cycles (Bond et al. 1993). Associated with these cycles are 'Heinrich Events', i.e. events of sudden iceberg discharge into the North Atlantic (Heinrich 1988; Bond et al. 1993). The origin and cause of these global teleconnected patterns is still unknown, even if a large proportion of the cooling in Europe and northern Asia during Heinrich Events is a meteorological response to cold surface water in the North Atlantic resulting from the surge of the Laurentian and Scandinavian ice sheets. The thermohaline circulation that originates in the North Atlantic and southern Arctic is a major force that drives not only the oceanic circulation but also regulates the global climate (Fig. 2). This «ocean current-conveyor belt» can switch between a normal mode of operation (formation of North Atlantic Deep Water throughsinking of the warm saline waters and releasing the heat energy) or be shut down by increased freshwater flux to the North Atlantic from the surge of the ice sheets (by reducing the salinity) triggering cold-wave conditions over Europe. Because the thermohaline conveyor belt straddles the globe, the warm/cold wave conditions can be transmitted across. This could possibly explain the global imprints of the D-O oscillations and the Heinrich events.

Proxy records of climate change from the northern Indian Ocean (Arabian Sea and Bay of Bengal) as well as from the Speleothem in China, India etc. suggest that the abrupt D-O and Heinrich-style events known from Greenland and the northern North Atlantic are a significant component also of low-latitude climate variability during the past 110,000 years, due to millennial to centennial fluctuations in the intensity of southwest monsoonal circulation (Schulz et al. 1998; Tiwari et al. 2011).

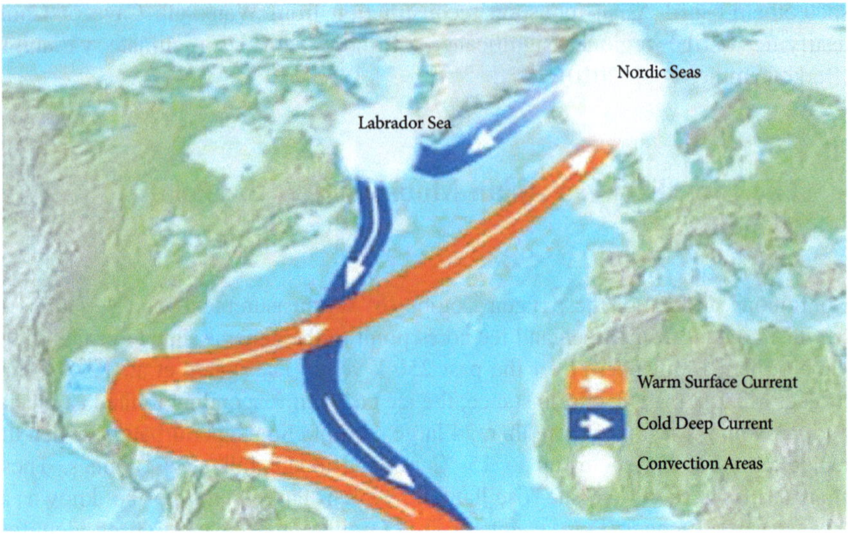

Fig. 2 North Atlantic thermohaline circulation. *Source* Adapted from UK Department of Environment, Food and Rural Affairs, Met Office Hadley Centre (2005)

These rapid fluctuations point to large-scale climate elements in the ocean-atmosphere system interacting rapidly in the high and low latitudes.

High-resolution Asian and Indian Summer palaeo-monsoon data derived from various natural archives such as tree-rings, speleothems and deep sea sediments reveal the persistence of the millennial-to centennial scale abrupt climate oscillations during the Holocene as well (Bhattacharyya and Yadav 1999; Yadava and Ramesh 1999; Sarkar et al. 2000; Luckge et al. 2001; Anderson et al. 2002; Burns et al. 2003; Fleitmann et al. 2003; Tiwari et al. 2011; Hong et al. 2014). Several of these abrupt changes are correlatable with the North Atlantic climate records (Schulz et al. 1998; Kudrass et al. 2001). For instance, speleothem records from China indicate the links between Asian monsoon and solar input and North Atlantic ice-rafting events, during the Holocene (Wang 2005). Similarly, monsoon proxy record from the Arabian Sea reveals several intervals of weak Indian summer monsoon that coincide with cold periods documented in the North Atlantic region (Gupta et al. 2003). Observation and modelling studies also show that the short-term multi decadal variations of the North Atlantic SST, i.e., the Atlantic Multi decadal Oscillation (AMO have corresponding variations in the Indian summer monsoon (Goswami et al. 2006; Lu et al. 2006). During the warm (cold) phase of AMO, the monsoon rainfall becomes abundant (scarce). These results indicate that the North Atlantic SST variations have persistent effects on the Indian summer monsoon rainfall at various time scales (Feng and Hu 2008). The proxy records of climate change from Antarctica also show the presence of abrupt short-term climate variabilities that can probably be tied to the annual/decadal changes in the Indian monsoon. For instance, correlation of Antarctic sea-ice anomaly in winter with monsoon rainfall over India during the period 1979–2000 seem to indicate that deficient monsoon years are preceded by more than normal sea-ice extent and vice versa (Dugam and Kakade 2004).

5 Summing Up

Proxy records of climate change clearly demonstrate that the earth's climate history spanning the last 100,000 years and beyond is marked by centennial-to millennial scale abrupt climate changes. However, a major missing link in our understanding of this phenomenon is knowledge of the forcing functions behind these abrupt changes as well as their spatial and temporal characteristics. To predict the evolution of the inter-regional climate linkages on shorter time scales, it is crucial to understand how they evolved in the past. This is especially significant as regards our predictive capabilities of the monsoon, considering that the regional monsoon system affects more than half of humanity worldwide. The credibility of future projections of monsoon behaviour is however, limited by the large spread in the simulated magnitude of precipitation changes and poor model agreements. The uncertainties can be reduced to an extent by focusing on constraints provided by quantitative records of long-term climate trends and shorter-term variability over the last 10 k years. Proxy records of climate change from a variety of sources demonstrate the persistence of the millennial-to centennial scale abrupt climate oscillations during the last glacial period. Although more subdued, abrupt changes in the Asian and Indian Summer monsoon that may be linked with episodic warming/cooling of the North Atlantic are also clearly discernible in the Holocene monsoon proxy records from the speleothems and Indian Ocean. Observation and modelling studies also show that shorter-term multi-decadal variations of the North Atlantic have corresponding variations in the Indian summer monsoon. Despite such a plethora of records, the causative factors behind such abrupt millennial-to shorter term climate changes remain topics of intense academic debate.

Acknowledgements I would like to express my gratitude to the organizers of the Third edition of SaGAA and in particular to Ms. Sulagna Chattopadhyay, not only for giving me an opportunity to present this paper, but also for the excellent hospitality extended during the Conference. The SaGAA Conferences have always remained inspirational and I have been fortunate to have been invited to make presentations in all, till date! I would also like to thank Dr. Rasik Ravindra, the Chairman of the Organising Committee for so graciously putting up with the inordinate and inexcusable delay on my part in submitting this paper. I hope the paper justifies the delay.

References

Altabet MA, Higginson MJ, Murray DW (2002) The effect of millennial-scale changes in Arabian Sea denitrification on atmospheric CO_2. Nature 415:159–162

Anderson DM, Overpeck JT, Gupta AK (2002) Increase in the Asian SW monsoon during the past four centuries. Science 297:596–599

Behl RJ, Kennett JP (1996) Brief interstadial events in the Santa Barbara basin, NE Pacific during the last 60 kyr. Nature 379:243–246

Berkelhammer M, Sinha A, Mudelsee M, Cannariato KG (2010) Persistent multi decadal power in the Indian summer monsoon. Earth Planet Sci Lett 290:166–172

Bhattacharyya A, Yadav R (1999) Climatic reconstruction using tree-ring data from tropical and temperate regions of India—a review. IAWA J 20:311–316

Bond G, Broecker WS et al (1993) Correlations between climate records from North Atlantic Sediments and Greenland ice. Nature 365:143–147

Broecker WS (1994) Massive iceberg discharges as triggers for global climate change. Nature 372:421–424

Broecker WS (1995) The glacial world according to wally, 2nd edn. Eldigio Press, Lamont-Doherty Earth Observatory of Columbia University. Palisades, New York

Burns SJ, Fleitmann D, Matter A, Kramers J, Al-Subbary AA (2003) Indian Ocean climate and an absolute chronology over Dansgaard/Oeschger events 9 to 13. Science 301:1365–1367

Charles CD, Lynch-Stieglitz J, Ninnemann US, Fairbanks RG (1996) Climate connections between the Hemisphere revealed by deep sea sediment core/ice core correlations. Earth Planet Sci Lett 142:19–27

Chawchai S (2014) Paleoenvironmental and paleoclimatic changes in northeast Thailand during the Holocene. Ph.D. Thesis, Stockholm University, Sweden

Dansgaard W et al (1993) Evidence for general climate instability of past climate from a 250-kyr ice core record. Nature 364:218–220

Dugam SS, Kakade SB (2004) Antarctica sea-ice and monsoon variability. Indian J Radio Space Phys 33:306–309

Feng S, Hu Q (2008) How the North Atlantic multidecadal oscillation may have influenced the Indian summer monsoon during the past two millennia. Geophys Res Lett 35:L01707. doi:10.01029/2007GL032484

Fleitmann D, Burns S, Mudelsee M, Neff U, Kramers Mangini A, Matter A (2003) Holocene forcing of the Indian monsoon recorded in a stalagmite from southern Oman. Science 300:1737–1739

Goswami BN, Madhusoodanan MS, Neema CP, Sengupta D (2006) A physical mechanism for North Atlantic SST Influence on the Indian summer monsoon. Geophys Res Lett 33:L02706

Greenland Ice-Core Project (Grip) Members (1993) Climate Instability during the last interglacial period recorded in the GRIP ice core. Nature 364:203–207

Gupta AK, Anderson DM, Overpeck JT (2003) Abrupt changes in the Asian southwest monsoon during the Holocene and their links to the North Atlantic Ocean. Nature 421:354–357

Heinrich H (1988) Origin and consequences of cyclic ice rafting in the northeast Atlantic Ocean during the past 130,000 years. Quat Res 29:142–152 doi:10.1016/0033-5894(88)90057-9

Hong B et al (2014) Abrupt variations of Indian and East Asian summer monsoons during the last deglacial stadial and interstadial. Quat Sci Rev 97:58–70

Keigwin LD, Curry WB, Lehman SJ, Johnsen S (1994) The role of the deep ocean in North Atlantic climate change between 70 and 130 kyr Ago. Nature 371:323–326

Kudrass HR, Hofmann A, Doose H, Emeisand K, Erlenkeuser H (2001) Modulation and amplification of climatic changes in the Northern Hemisphere by the Indian summer monsoon during the past 80 ky. Geology 29:63–66

Kutzbach JE (1981) Monsoon climate of the early Holocene: climate experiment with the earth's orbital parameters for 9000 years ago. Science 214:59–61

Laskar J, Robutel P, Joutel F, Gastineau M., Correia ACM, And Levrard B (2004) A long-term numerical solution for the insolation quantities of the earth. Astron Astrophys 428:261–285

Lu R, Dongand B, Ding H (2006) Impact of the Atlantic multi decadal oscillation on the Asian summer monsoon. Geophys Res Lett 33:L24701

Luckge A, Doose-Rolinski A, Khan AA, Schulz H, von Rad U (2001) Monsoonal variability in the northeastern Arabian Sea during the past 5000 Years: geochemical evidence from laminated sediments. Palaeogeogr Palaeoclimatol Palaeoecol 167:273–286

Marzin C, Kallel N, Kageyama M, Duplessy JC, Braconnot P (2013) Glacial fluctuations of the Indian Monsoon and their relationship with North Atlantic climate: new data and modelling experiments. Clim Past 9:2135–2151

Overpeck J (1996) Warm climate surprises. Science 271:1820–1821

OverpeckJ T, Anderson DM, Trumbore S, Prell WL (1996) The southwest monsoon over the last 18,000 years. Clim Dyn 12:213–225

Sarkar A, Ramesh R, Somayajulu BLK, Agnihotri R, Jull AJT, Burr GS (2000) High resolution Holocene monsoon record from the eastern Arabian Sea. Earth Planet Sci Lett 177:209–218

Schulz H, Radand U, Erlenkeuser H (1998) Correlations between Arabian Sea and Greenland climate oscillations of the past 110,00 years. Nature 393:54–57

Sirocko F, Leuschner D, Staubwasser M, Maley J, Heusser L. (1999) In: Clark P, Webb RS, Keigwin LD (eds) Mechanisms of global climate change at millennial time scales. American Geophysical Union, Washington D.C., pp 113–126

Tiwari M, Singh AK, Ramesh R (2011) High-resolution monsoon records since last glacial maximum: a comparison of marine and terrestrial paleoarchives from south Asia. J Geol Res 2011:1–12

Wang Y (2005) The Holocene Asian monsoon: links to solar changes and North Atlantic climate. Science 308:854–857

Wang YJ, Cheng H, Edwards RL, An ZS, Wu JY, Shen CC, Dorale JA (2001) A high-resolution absolute-dated late pleistocene monsoon record from Hulu Cave, China. Science 294: 2345–2348

Yadava MG, Ramesh R (1999) Paleomonsoon record of the last 3400 years from speleothems of tropical India. Gondwana Geol Mag 4:141–156

Yanase W, Abe-Ouchi A (2007) The LGM surface climate and atmospheric circulation over East Asia and the North Pacific in the PMIP2 coupled model simulations. Clim Past 3:439–451

Part III
Global Climate Change and the Himalayan Region

Inventory and Monitoring of Snow and Glaciers of the Himalaya Using Space Data

Ajai

Abstract The 2400 km long Himalayan mountain chain and the glaciers therein, contribute to most of the perennial rivers that constitute the life line for northern plains of India and adjoining areas because of their potential for irrigation, hydropower generation and being critical indicators of climate change. A large number of these glaciers, especially those in the Indus, the Ganga and the Brahmaputra basins, have been monitored for more than last two decades, using different satellite data. Space Application Centre of ISRO has prepared an updated inventory of Himalayan glaciers on 1:50,000 scale using Resourcesat-1 satellite data (2004–07), which serves an important document for monitoring the health of these glaciers. This inventory has shown presence of a total number of 32,392 glaciers, of different sizes, in the three above mentioned basins. The monitoring of snout of these glaciers in the two periods of 1989–90 to 2001–04 and 2001–02 to 2010–11 has presented a contrasting data. During the period of 1989–90 to 2001–2004, 76% of the glaciers have shown retreat, 7% have advanced and 17% have shown no change. As compared to this during the next decade i.e. 2001–02 to 2010–11, only 12.3% glaciers have shown retreat, 86.6% of glaciers have shown stable front and 0.9% have shown advancement. Though the space borne satellite data has been used for covering large tracks of inaccessible terrain of Himalaya for snout monitoring, there is need to include and develop models to estimate the thickness of glaciers using this space data. The data from space borne Lidar and interferometric SAR, used in association of GPR survey data and ground truthing, can help in estimation of thickness of glacier ice and mass balance an important parameter that is required to be added in the monitoring.

Keywords Himalayan glacier · Inventory · Monitoring · Satellite images

Ajai (✉)
ES, CSIR, Space Applications Centre (ISRO), Ahmedabad, India
e-mail: drajai_in@yahoo.com

P.S. Goel et al. (eds.), *Science and Geopolitics of The White World*,
DOI 10.1007/978-3-319-57765-4_8

1 Introduction

Snow and glaciers are vital to human beings as they play a critical role in making our river perennial, controlling global and regional climate, and serve as a sensitive indicator of climate change. Permanent, seasonal, and temporary are the three components of snow cover. Temporary and seasonal snow-cover occurs in winters while permanent snow cover is retained for many years. Permanent snow cover occurs, mainly, in Antarctica, Greenland and above permanent snow line in mountainous areas. Snow melt runoff is important for water storage for drinking, irrigation, and hydroelectric power generation and to maintain the ecosystems. Another important climatological role of snow cover is the thermal insulation provided by it which reduces the exchange of heat between the ground and the atmosphere.

About 16 million km^2 i.e. 10% of the earth's land area is covered by ice and the volume of the ice is about 26 million km^3 (Flint 1964; Benn and Evans 2010). The maximum ice cover has been estimated in Antarctica region which is nearly 13.5 million km^2. Greenland (1.74 million km^2) is the next highest ice covered region. The remaining icy regions are the mountain glaciers which are mainly valley glaciers. Mountain glaciers are in the Himalayas, Alps, Andes, Rockies, China, Russia, Africa, Alaska, and New Zealand.

Among all the mountainous regions of the world, glaciers of the Himalaya constitute the largest concentration of freshwater reserve outside the polar region. Thus, it is also called as the "Third Pole". Snow and glacier ice melt from these frozen reservoirs makes the rivers, originating from the Himalaya, perennial and helped in flourishing the civilizations along the banks of these rivers from time immemorial. The Himalayan mountain ranges extend from Kashmir in the west to Arunachal Pradesh in the east. In India, the Himalayas occupy the parts of Jammu & Kashmir, Himachal Pradesh, Uttarakhand in West and state of Sikkim, Assam and Arunachal Pradesh in the east. In between, the two countries Nepal and Bhutan also link the Himalayan chain. Hydrologically, the Himalayas are drained by the three major river systems namely the Ganges, the Indus and the Brahmaputra. The Ganges originates in Uttarakhand region of India and flows down to the Bay of Bengal. Its major tributaries are Bhagirathi, Alaknanda, Yamuna and Kosi rivers. The later originates in the Nepal Himalayas. The Brahmaputra flows east ward across Tibetan plateau and drains to the Bay of Bengal. Its major tributary is Tista River which drains through the Sikkim state. These rivers are perennial and bring water to the northern plane of the country. The Indus flows to the Arabian Sea and its major tributaries are Satluj, Chenab and Jhelum rivers. The rivers draining out of the Himalayas have made the frontal portions of the Himalayas into a very vast alluvial plain that is one of the most fertile regions of our country.

2 Characteristics of Himalayan Glaciers

The Himalayas with a length of about 2400 km have arcuate shape and therefore occupy a wide range of latitudinal variations. From north to south, the Himalayas can be divided into the Great Himalayan ranges, Lesser Himalayan ranges and Shivalik ranges. These ranges characterize high to low altitudes from north to south respectively. The altitudinal and latitudinal variations provide different temperature regimes to these areas. Western part of the Himalayas is influenced by western monsoons where as the eastern Himalayan region is influenced primarily by SE monsoon from Bay of Bengal. Large areas in the Himalayas are covered by snow during winter season. Area of snow can change significantly during winter and spring. The onset of seasonal snowfall in Indian Himalaya begins in Kashmir in late September/early October months and slowly shifts towards lower latitudes. The melting season of snow and glacier also progresses similarly from months of July to end of September. The accumulation and ablation time in the eastern Himalayan region (Sikkim and Arunachal Pradesh) is little late by about three months. Altitude of lowest snow line also significantly varies from Kashmir, Uttarakhand, Nepal, Sikkim and Bhutan. The snow line comes down to an altitude of approximately 2500 m in winters. The snow line at the end of ablation varies between 4500 to 5000 m.

Increase in concentration of glaciers in the Himalaya varies from northwest to northeast according to the variation in altitude and latitude of the region. Siachin glacier in Kashmir, Gangotri glacier in Uttarakhand (Ganga basin), Bara Shigri glacier in Himachal (Chenab), Baltoro glacier in Karakoram (Indus basin) and Zemu glacier in Sikkim (Tista) are a few famous glaciers of the Himalayas. The glaciers in the Himalaya have high relief and occur above 4000 m above mean sea level (amsl) up to peak of Mt. Everest 8848 m amsl. These glaciers are temperate glaciers in nature. Many glaciers are debris covered in their ablation zone. Thus these glaciers appear dark as compared to the mountain glacier of higher latitude. Debris cover reduces the glacier surface albedo.

Due to high relief, the Himalayan glaciers cover large altitudinal range as compared to the glaciers in other non-polar mountainous ranges. The large altitudinal range provides different climatic zones and wind zones to these glaciers. This has implications on the accumulation and ablation patterns of snow and ice if there is a rise in temperature.

3 Inventory of Himalayan Glaciers

The glacier inventory primarily deals with the occurrence and distribution of glaciers and also provides details for each glacier on morphology, dimensions, orientation, elevation, etc. for both the active glacier component as well as the associated de-glaciated valley. Satellite images have been extensively used in studying and mapping of the glaciers (Ostrem 1975; Dozier 1984; Hall et al. 1988;

Bolch et al. 2010; Ajai et al. 2011; Hewitt 2014). In India, first glacier inventory of the Indian Himalaya using images available from Landsat and IRS 1A & 1B satellites was carried out at 1:250,000 scales (Kulkarni 1992a, b). Detailed glacier inventory for Sutlej basin at 1:50,000 scales was followed up (Kulkarni et al. 1999). The glacier features, which are mapped using satellite data, are glacier boundaries; accumulation zone, the snow line, the ablation zone, deglaciated valleys and moraine dammed lakes (Ajai et al. 2011). The inventory of Himalyan glaciers have also been produced by Geological Survey of India (Kaul et al. 1999; Sangewar and Shukla 2009)

Inventory of the glaciers of the entire Himalaya (the Indus, Ganga and Brahmaputra basins) have been carried out on 1:50,000 scale using IRS P-6 AWiFS data (Sharma et al. 2013). Figure 1 shows the spatial extent of the present glacier inventory as depicted on the satellite image (False Color Composite). The inventory includes the complete Ganga, Brahmaputra and the Indus basin (only a small portion in the North West is not included). This inventory includes even the glaciers of these three basins which are located outside the Indian Territory but draining into India (Sharma et al. 2013).

The glacier inventory map depicts the glaciers and their spatial distribution. The significant glacier morphological features for each of the glacier are mapped and appropriately represented on the map by a pre-defined colour scheme. The mapped glacier features comprise of glacier boundary with separate accumulation area and

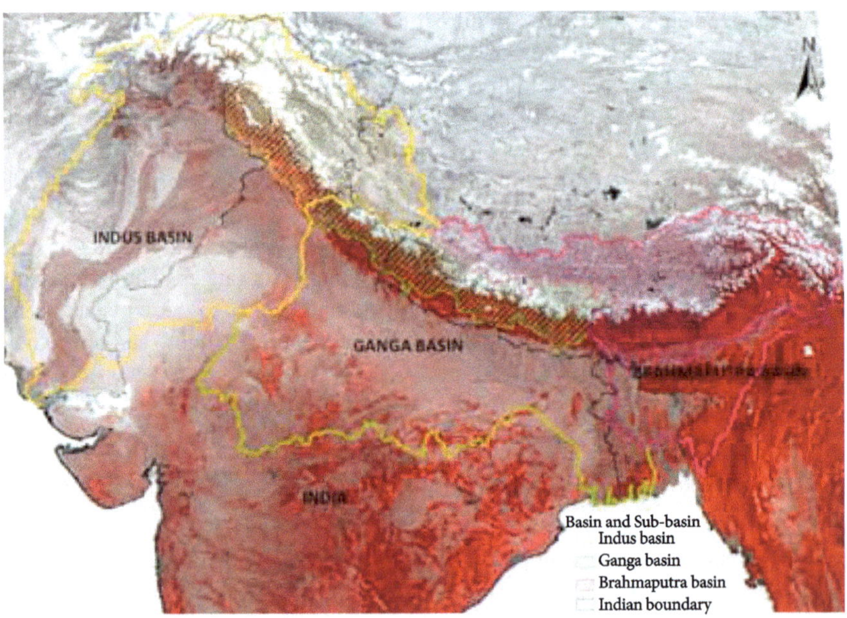

Note: Satellite Image (False Colour Composite) of Snow Cover/Glaciated Regions of the Himalayas along with basin boundaries of the Indus, Ganga and Brahmaputra. Hatching shows the areas covered for glacier inventory.

Fig. 1 The spatial extent of the present glacier inventory. *Source* Satellite Image, ISRO ablation area. The ablation area is further divided into ablation area ice exposed and ablation area debris covered. The Moraines like median, lateral and terminal moraines present on the glacier are separately mapped and delineated. The supra-glacier lakes occurring on the glaciers are also delineated. The snout is marked as a point location depicting the end of the glacier tongue. The de-glaciated valley associated with the glacier is also delineated along with the associated moraines both lateral and terminal moraines and the moraine dam lakes.

The results of the inventory of glaciers, providing details on the number and area of the glacier features mapped for Indus, Ganga and Brahmaputra basins are given in Table 1. Specific measurements of mapped glaciers features is the input for generating the glacier inventory data sheet with 37 parameters as per the UNESCO/TTS standards and 11 additional features associated with the deglaciated valley. The data sheet provides glacier wise details, mainly related to the glacier identification, in terms of number, name, glacier locations, information on the elevation, dimensions and orientation etc.

Total number of glaciers and the glaciated area in these three basins put together are 32,392 and 71,182.08 km^2 respectively (Sharma et al. 2013; Ajai et al. 2011). The Indus basin has 16,049 glaciers having glaciated area of 32,246.43 km^2. The Ganga basin has 6237 glaciers occupying 18,392.90 km^2 of glaciated area. There are 7 glaciated sub-basins in Ganga basin. The Brahmaputra basin has 10,106 glaciers occupying 20,542.75 km^2 of glaciated area. There are 27 sub-basins in Brahmaputra basin covering glaciated area. The total glaciated area and the number

Table 1 Details of the glacier inventory of the Himalayas

S No.	Basin	Indus	Ganga	Brahmaputra	All basins total
	Characteristics	Area in km^2	Area in km^2	Area in km^2	Area in km^2
1	Sub-basins (Nos.)	18	7	27	52
2	Accumulation area	19,265.98	10,884.60	12,126.36	42,276.94
3	Ablation area debris	6650.95	4844.70	5264.90	16,760.55
4	Ablation area ice exposed	6310.58	2663.50	3081.48	12,055.56
5	Total no. of glaciers	16,049	6237	10,106	32,392
6	Total glaciated area	32,246.43	18,392.90	20,542.75	71,182.08
7	No. of Permanent Snow fields and Glacierets	5117	641	3651	9409
8	Area under Permanent Snow fields and Glacierets	991.68	198.70	1282.92	2473.30
9	No. of Supra-glacier lakes	411	87	474	972
10	Area of Supra-glacier lakes	18.92	15.20	70.01	104.13
11	No. of Moraine dam/Glacial lakes	469	194	226	889
12	Area of Moraine dam/Glacial lakes	33.82	64.10	70.15	168.07

Source Sharma et al. (2013)

of glaciers in the Indian Himalaya are 37,266 and 16,445 km^2 respectively (Sharma et al. 2013).

Results of this inventory show that the accumulation and ablation areas of the glaciers are the largest in the Indus basin. The ratio of accumulation to ablation area is high in the Brahmaputra basin. This indicates that the glaciers of the Indus basin are having larger feed area and hence are relatively stable as compared to the other two basins (Sharma et al. 2013). Indus basin has 51.3% of the ablation area covered with debris whereas the Ganga and the Brahmaputra basins have 64.5 and 63.1% debris cover respectively. All the three basins put together, about 60% of the glaciated area in the ablation zone is covered with debris. The percentage of debris covered glaciers is highest in the Ganga basin. Though the number of glacial lake/moraine dammed lake is higher in Indus basin but the total area of such lakes is the lowest among these three basins which means that the lakes are smaller in size in Indus basin and larger in the Ganga and the Brahmaputra basin. The number of supra glacier lakes is 474 (with an area of 70.01 km^2) in the Brahmaputra basin which is largest amongst all the three basins. The area of permanent snow field is smallest in Ganga basin which is hardly 198.7 km^2. The area under permanent snow fields are 991.68 and 1282.92 km^2 in the Indus and the Brahmaputra basins respectively. Glaciers of all the basins have been classified in different classes zones based on the size. The number of glaciers having size less than 1 km^2 is highest in Indus basin and lowest in the Ganga basin. Ganga basin has maximum number of large sized glaciers as compared to the Indus and Brahmaputra basins (Fig. 2).

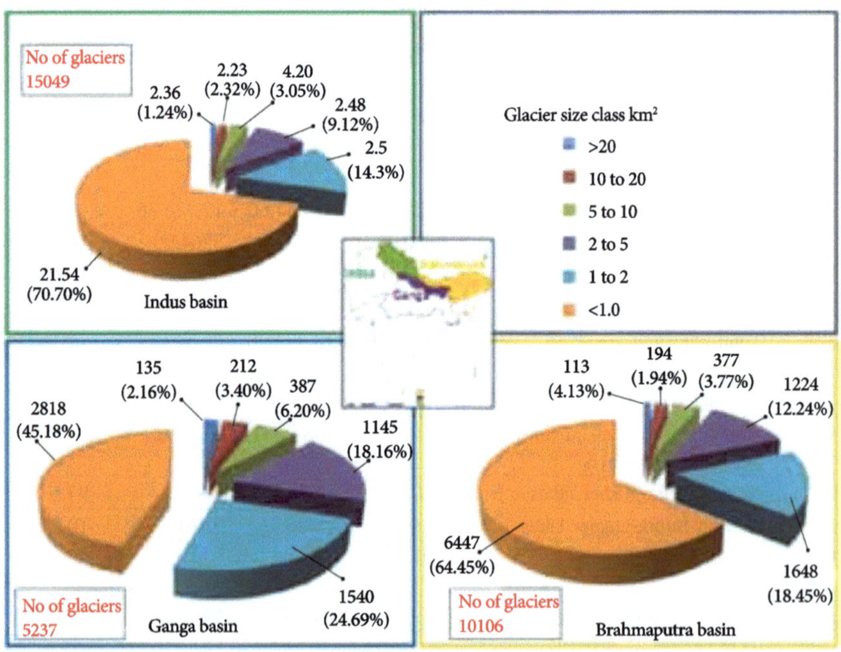

Fig. 2 Glacier size distributions in the Indus, Ganga and Brahmaputra basins. *Source* Kasturirangan et al. (2011)

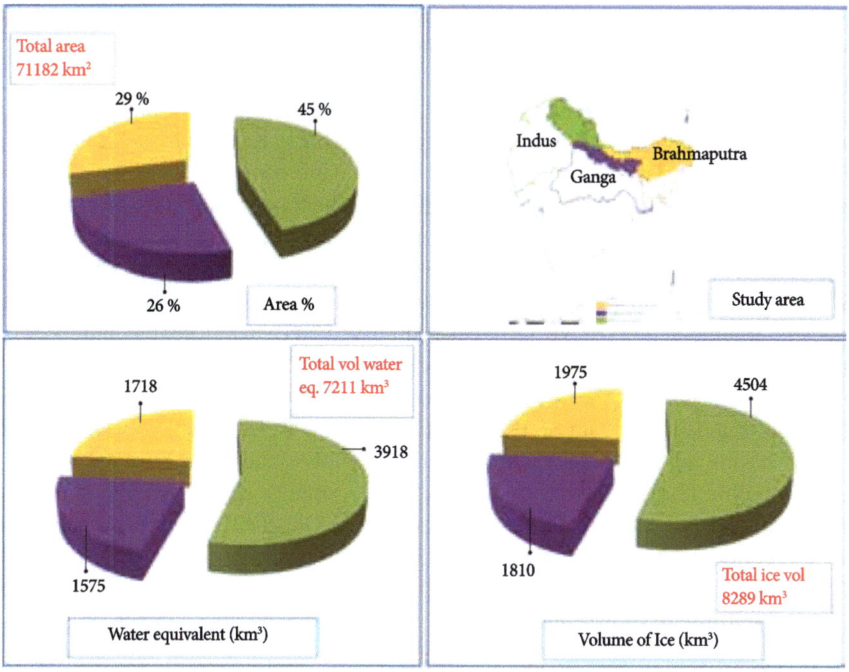

Fig. 3 Indus, Ganga and Brahmaputra basins area and volume estimation. *Source* Kasturirangan et al. (2011)

Computing the glacier ice volume and water equivalent (stored in frozen condition) is important from the water security point of view. Volume of ice has been computed for the three basins using glacier area and ice thickness (Chaohai and Sharma 1988). Figure 3 shows the glaciated area, glacier volume and water equivalent for the three basins. Indus has the largest ice volume and the water stored.

4 Monitoring Retreat/Advance of Glaciers

The retreat or advance of individual glaciers depends upon the variations in mass balance which in turn is governed by static and dynamic factors. The static factors are latitude, slope, orientation, size of the glaciers, altitude of the snout, the debris cover and altitudinal distribution of glaciers. The dynamic parameters are the ratio of the accumulation area to the ablation area, variations in temperature, precipitation, heat flow from earth crust and cloud cover (Brahmabhatt et al. 2012). The other most important process responsible for the retreat or advance of glaciers is the energy balance on the glacier surface. Studies have been carried out to model and compute the energy balance on the Gangotri and Chhota Shigri glaciers and the impact of the change in the energy balance on the melting of glacier ice (Rastogi and Ajai 2013a, b).

Advance/retreat of glaciers has been monitored in fourteen sub-basins using multi-date data (Bahuguna et al. 2010; Ajai et al. 2011; Kasturirangan et al. 2011). These basins are selected to represent all the climatic zones of the Himalayas. For long term change monitoring, Survey of India (SOI) topographical maps of 1962 at 1:50,000 scales have been used as reference for glacial extent. The long term (40 years) retreat/advance has been monitored using glacier boundaries from 1962 SOI topographical maps and the satellite data of 2001–04 time frame. In some, glaciers fragmentations have also been observed and therefore the numbers of glaciers in a specific range of area has also changed. Retreat/advance has also been computed based on glacier extent mapped from IRS data available from 2001 to 2004 and Landsat TM data of 1989–90 time frames. Results are given in Table 2.

The basins having larger sized glaciers e.g. Nubra and Tista have shown very less/negligible retreat whereas the basins having smaller size of glaciers e.g. Spiti and Zanskar have shown relatively higher retreat. Out of 2190 glaciers monitored for the period 1989–90 to 2001–04, 1673 (76%) glaciers have shown retreat, 158 (7%) glaciers have advanced, whereas 359 (17%) glaciers have shown no change (Table 3). Table 4 shows the details in terms of the basins, number of glaciers

Table 2 Mean annual loss/gain in glaciated area in each basin: based on SOI maps and satellite images

Basin	Mean glacier area (sq km)	SOI maps 1962/1969-Satellite images 2001/2004		Satellite images 1989/1990–2001/2004	
		No. of glaciers monitored	Mean annual loss in area (%)	No. of glaciers monitored	Mean annual loss in area (%)
Chandra	6	116	0.51	3	0.23
Bhaga	3.27	111	0.77	10	0.19
Parbati	5.48	90	0.51	10	0.83
Warwan	3.2	230	0.46	180	0.17
Bhut	3.14	143	0.18	28	0.46
Alaknanda	3.82	274	0.33	119	0.67
Bhagirathi	6.5	183	0.26	153	0.11
Dhauliganga	4.125	104	0.37	–	–
Gauriganga	9.38	–	–	29	0.27
Suru	2.65	215	0.49	355	0.82
Zanskar	1.75	631	0.38	463	1.80
Spiti	1.41	337	0.41	722	2.23
Tista	9	–	–	34	0.07
Nubra	37	31	0.15	84	0.00

Source Ajai et al. (2011)

Table 3 Number of glaciers showing advance, retreat and no change (based on the satellite images of 1989–90 and 2001/2004)

Basin	No.	Retreat	Advance	No change
Chandra	3	3	–	–
Bhaga	10	10	–	–
Warwan	180	32	–	148
Bhut	28	17	–	11
Alaknanda	119	119	–	–
Bhagirathi	153	44	6	103
Gauriganga	29	20	–	9
Suru	355	299	39	17
Zanskar	463	422	41	–
Parbati	10	10	–	–
Spiti	722	648	39	35
Tista	34	23	8	3
Nubra	84	26	25	33
Total	2190	1673 (76%)	158 (7%)	359 (17%)

Source Ajai et al. (2011)

Table 4 Loss/gain in area (km^2) of glaciers in different basins based on satellite images

Basin	No of Glaciers	Year	Area (sq km)	Year	Area (sq km)	Loss/Gain %
Chandra	3	1989	107	2002	104	3
Bhaga	10	1990	90	2001	88	2
Warwan	180	2001	513	2007	510	1
Bhut	28	1989	217	2002	203	6
Alaknanda	119	1990	393	2005	355	10
Bhagirathi	153	1989	867	2005	851	1.8
Gauriganga	29	1990	272	2005	261	4
Suru	355	1990	506	2001	459	9
Parbati	10	1998	113	2004	107	5
Tisat	34	1990	305	2004	301	1
Zanskar	463	2001	775	2006	709	9
Spiti	722	2001	718	2007	622	13.4
Nubra	84	1989	3159	2001	3163	0.0013*
Total	2190					

*Shows increase in glaciated area. For all other basins there is decrease in the glacier area during the period of monitoring

Source Ajai et al. (2011)

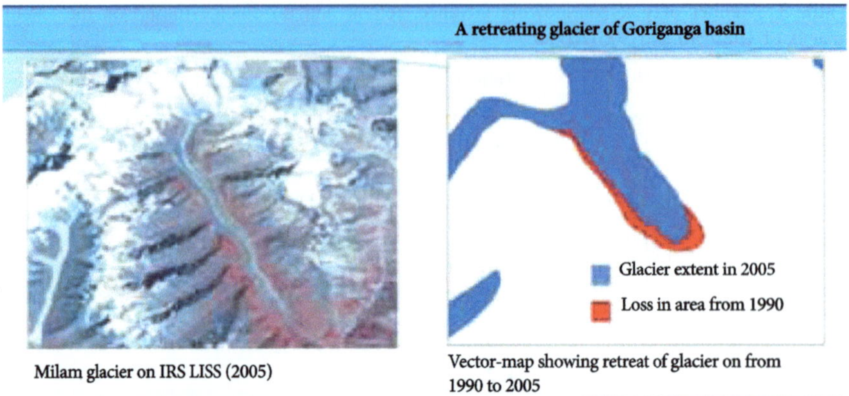

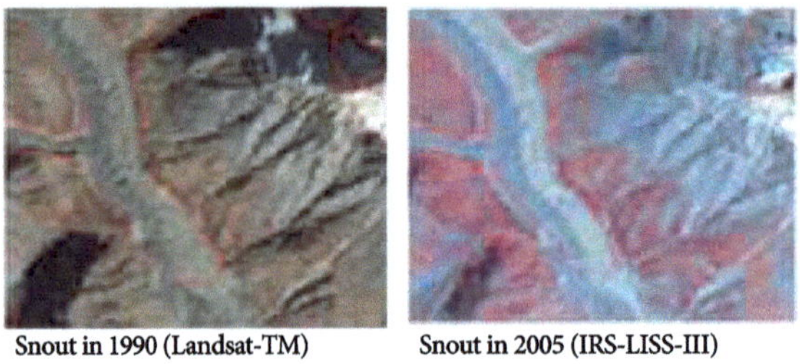

Area in 1990 (Sq. Km.)	68.77
Area in 2005 (Sq. Km.)	66.12
Change area (Sq. Km.)	2.65
Change in volume 1990 (water eq.)	0.538 Km^3

Fig. 4 Retreat of Milam glacier in Goriganga basin of Uttarakhand. *Source* Kasturirangan et al. (2011)

monitored in each basin, the years of monitoring and the average loss in glacier area. On the average, there is a loss of 3.75 % of glaciated area. Figures 4, 5 and 6 show the examples of retreating, static and advancing glaciers observed through use of Landsat and IRS data. Milam glacier in Goriganga basin has shown a retreat of 2.65 km^2 during the period 1990–2005 (Fig. 4). The glacier shown in Fig. 5 does not show any change in its snout position during the period of 1998–2007. Kichik Kumdum glacier of the Shyok basin (Indus basin) has shown an increase of 1.94 km^2 during the period 1989–2001 (Fig. 6). In order to validate the retreat/advance in the field, one glacier in each basin has been visited and snout position has been measured using high precision GPS.

Space applications Centre, ISRO in collaboration with seven other important organizations of the country has carried a study to find changes in the extent of the Himalayan glaciers during the decade 2001–2011 using satellite data (Bahuguna et al. 2014). A total of two thousand and eighteen glaciers representing climatically diverse terrain in the Himalaya were mapped and monitored using IRS LISS-III images of 2000/01/02 and 2010/11. It included glaciers of Karakoram, Himachal, Uttarakhand, Nepal and Sikkim regions. Among these, 1752 glaciers (86.8%) were

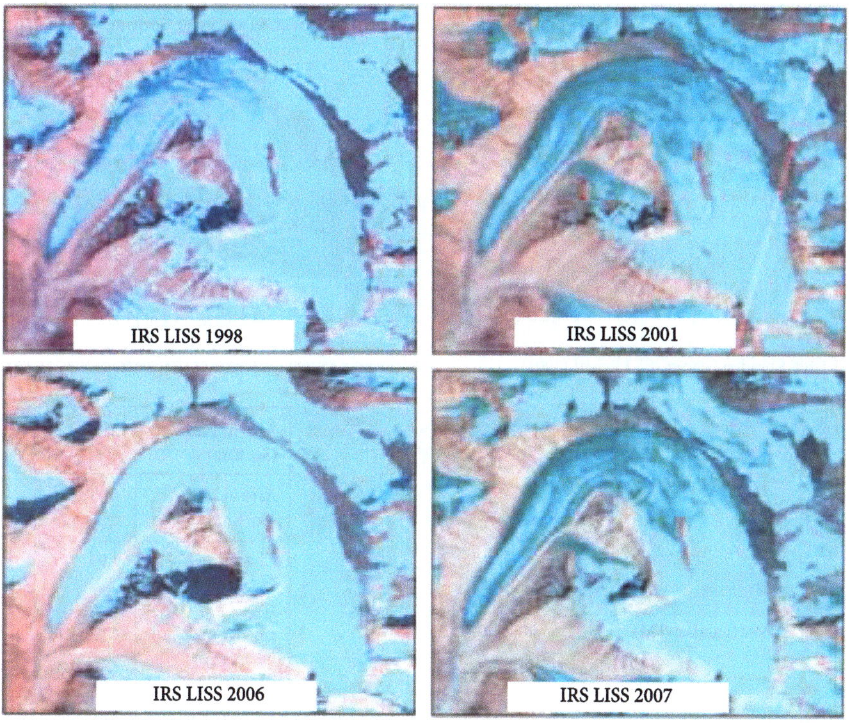

Fig. 5 Glaciers Showing No Change (Static) in Chandra Basin. *Source* Kasturirangan et al. (2011)

observed having stable fronts (no change in the snout position and area of ablation zone), 248 (12.3%) glaciers exhibited retreat and 18 (0.9%) of them exhibited advancement of snout. The net loss in 10,258.68 km^2 area of the 2018 glaciers put together was found to be 20.94 km^2 or 0.2% of the total glacial area. The details in terms of the number of glaciers showing retreat/advance or no change in the different regions are shown in Fig. 7.

It is important to note that during 1989/90–2001/02, about 76 % of the glaciers have retreated and only 17% have shown stable front (Table 3). Whereas during the next decade i.e., 2001/04–2010/11, the per cent number of glaciers having stable front (no change in snout position) has increased to 86.8%. The per cent number of glaciers showing retreat has come down to only 12.3% from 76% in the previous decade. On the average, there was a loss of 3.75% in the glaciated area during the period 1989/90–2001/02, whereas this loss has come down to only 0.2% during the decade 2001/04–2010/11. From the above, it is clear that the rate of retreat of the Himalayan glaciers has come down during the decade 2001–2010 as compared to the previous decade.

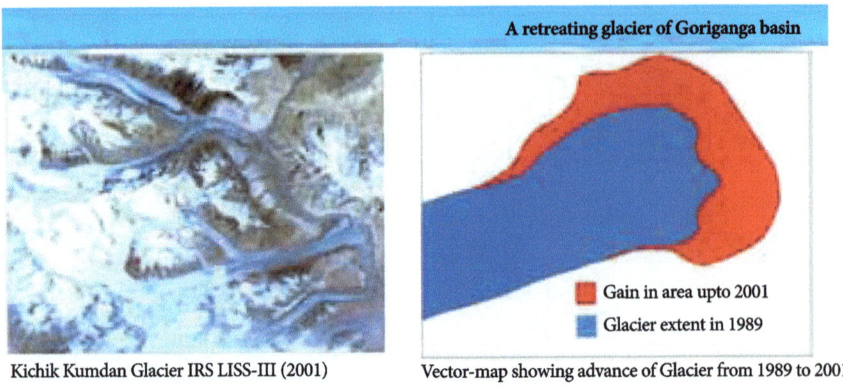

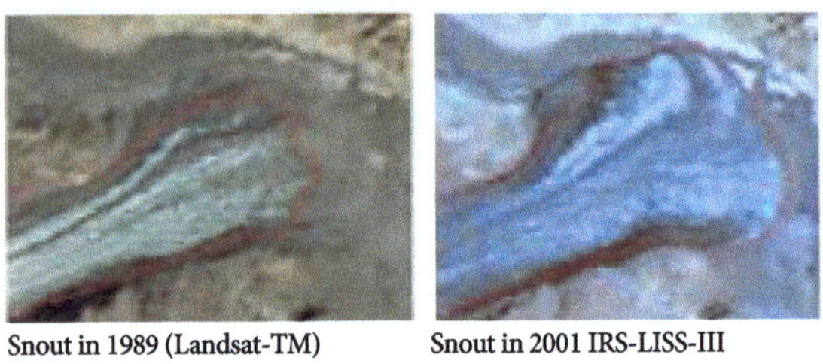

Area in 1989 (Sq. Km.)	65.44
Area in 2001 (Sq. Km.)	67.38
Change area (Sq. Km.)	1.94
Change in volume (water eq.)	0.450 Km^3

Fig. 6 Glaciers Showing No Change (Static) in Chandra Basin. *Source* Kasturirangan et al. (2011)

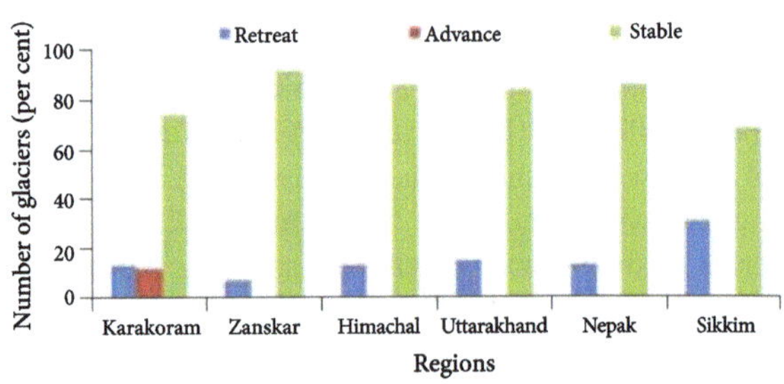

Fig. 7 Number of Glaciers showing retreat, advance or no change*. *Source* Bahuguna et al. (2014)

5 Glacier Mass Balance

The mass balance of the glacier is usually referred to as the total loss or gain in glacier mass at the end of the hydrological year. It is an important parameter required to understand the health of the glacier as well as availability of water over a long period of time. Glacial mass balance can also provide very vital clues for climatic change.

Mass balance is estimated by measuring the total accumulation of seasonal snow and ablation of snow and ice. The accumulation (input) includes all forms of deposition, precipitation mainly and ablation (output) means loss of snow and ice in the form of melting, evaporation and calving etc., from the glacier. The boundary between accumulation and ablation area is the Equilibrium line. The difference between net accumulation and net ablation for the whole glacier over a period of one year is net balance. The net balance for each glacier is different in amount and depends upon the size/shape of the glacier and climatic condition of the area. The net balance per unit area of glacier is specific mass balance, expressed in mm of water equivalent. There is wide variation in mass changes from time to time and place to place on the glacier due to the various factors.

There are various methods for estimating glacier mass balance. One of the methods using satellite data is based on AAR (accumulation area ratio) approach. A relationship between AAR and mass balance is developed using field mass balance data (Kulkarni 1992a, b). On the basis of accumulation area ratio (area of accumulation divided by whole area of glacier) mass balance in terms of gain or loss can be computed. Multi temporal AWiFS data from IRS P-6 satellite has been used to delineate the snow line and its shift during the ablation season. Finally, the snow line at the end of ablation season (the equilibrium line) is delineated to compute AAR.

The computed AAR is used in the 'AAR—Mass balance model' to compute the yearly mass balance. Figure 8 shows mass balance of glaciers for a few basins for the year 2005, 2006 and 2007. The AAR for most of the basins shows negative specific mass balance during the years 2005–2007. Figure 9 shows that mass balance for 700 glaciers in 10 basins, well distributed in the Himalayas, based on AAR approach. These glaciers are showing negative mass balance.

Results of the mass balance computed for the years 2010, 2011 and 2012 for a large number of glaciers from the ten basins of the Himalaya reveals that more than 50% of the glaciers in the year 2012 showed positive mass balance (Fig. 10). About 80% of the glaciers showed positive mass balance during the hydrological years 2010 and 2011. The trend has reversed during the period 2005–07 to 2010–2012. Majority of the glaciers have negative mass balance during 2005, 2006 and 2007

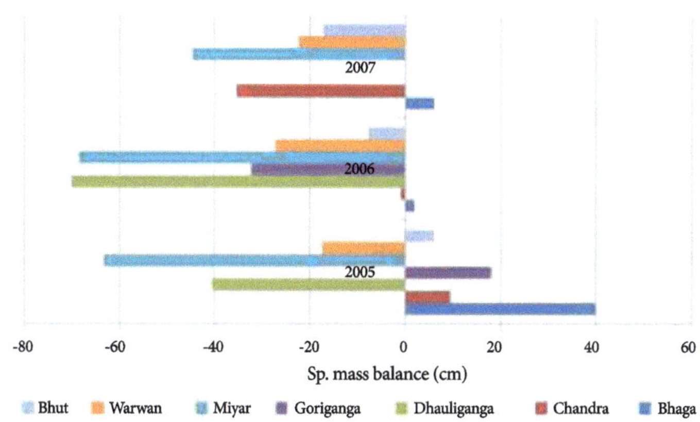

Fig. 8 Basin wise specific mass for the year 2005–07. *Source* Kasturirangan et al. (2011)

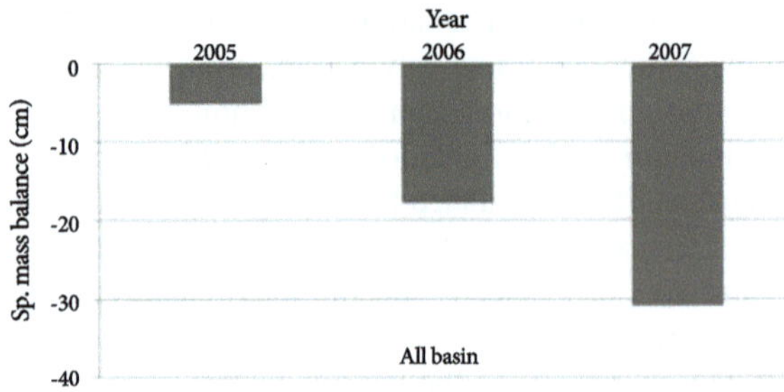

Fig. 9 Specific mass showing a negative trend from 2005 to 2007. *Source* Kasturirangan et al. (2011)

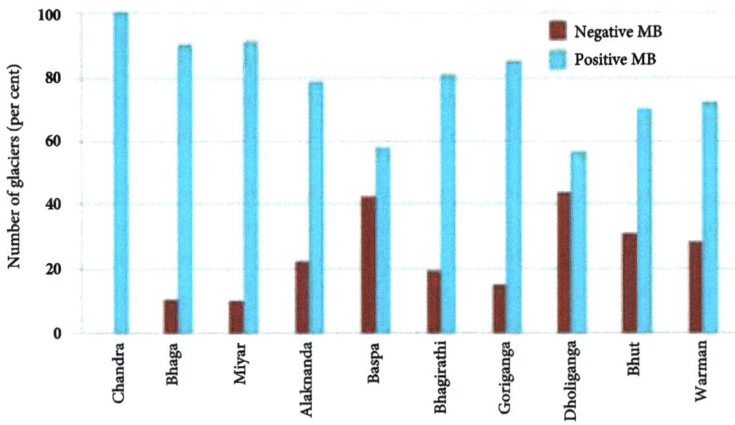

Year	2012	2011	2010
No. of Glaciers	1037	843	733
Negative	478	151	148
Positive	559	692	585

Fig. 10 Positive and Negative Mass Balance of Glaciers, 2010, 2011 and 2012. *Source* SAC (2015)

whereas the mass balance became positive for majority of glaciers during the years 2010, 2011 and 2012.

6 Monitoring of Snow Cover

Studying the snow cover pattern over time is very important for computing the snow melt runoff. Data from AWiFS sensor on board Resourcesat-1 satellite has been used to monitor seasonal snow cover of the Himalayas. Algorithm based on Normalized Difference Snow Index (NDSI) is used to map snow cover

(Dozier 1984, 1989; Kulkarni et al. 2006). NDSI is calculated using the ratio of green (band 2) and SWIR (band 5) channel of AWiFS sensor (Ajai et al. 2011). Snow cover products from AWiFS/MODIS have been used to monitor the snow cover pattern in the Himalayas.

The snow accumulation and ablations curves are different for these three basins, depending upon their locations and altitude distribution of the basin. Snow cover has been monitored over time for the Indus, Ganga and Brahmaputra basins. Detailed level monitoring of snow cover has also been done for a large number of sub-basins in the Himalayas. Snow cover for the months of September (the end of ablations) and March (end of accumulation) for the Indus, Ganga and Brahmaputra basins from the year 2000 to 2015 are given in Fig. 11. The snow cover in the Indus

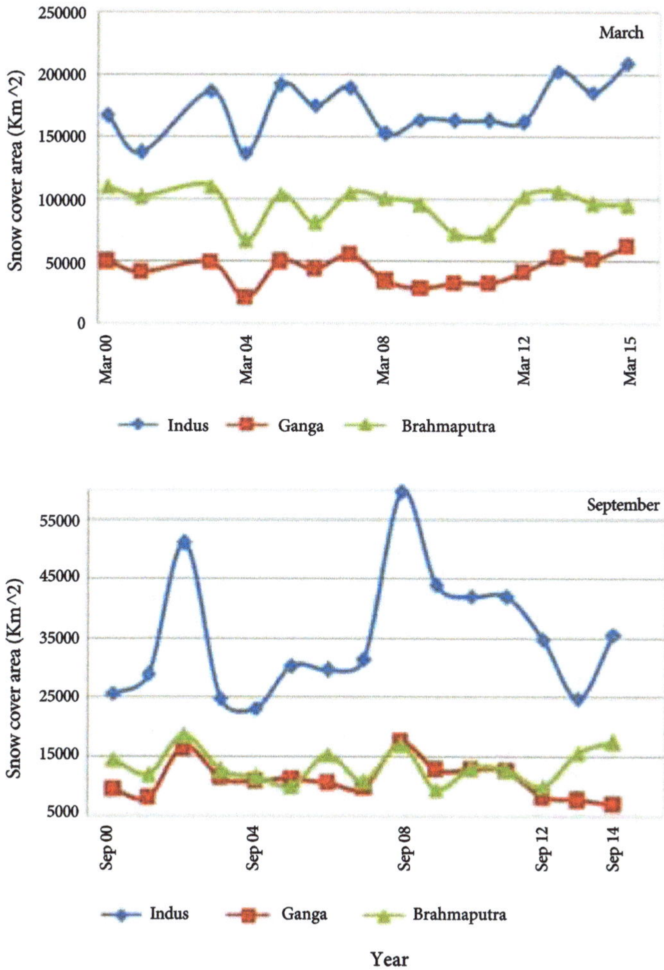

Fig. 11 Snow cover variability during the years 2000–2015 for three basins in September and March. *Source* Modified from Singh et al. (2014)

basin is relatively higher as compared to the Ganga and the Brahmaputra basins in both September and March. There is an increasing trend in the snow cover for Indus basin in the month of September, where as it does not show any trend for the Ganga and Brahmaputra basins.

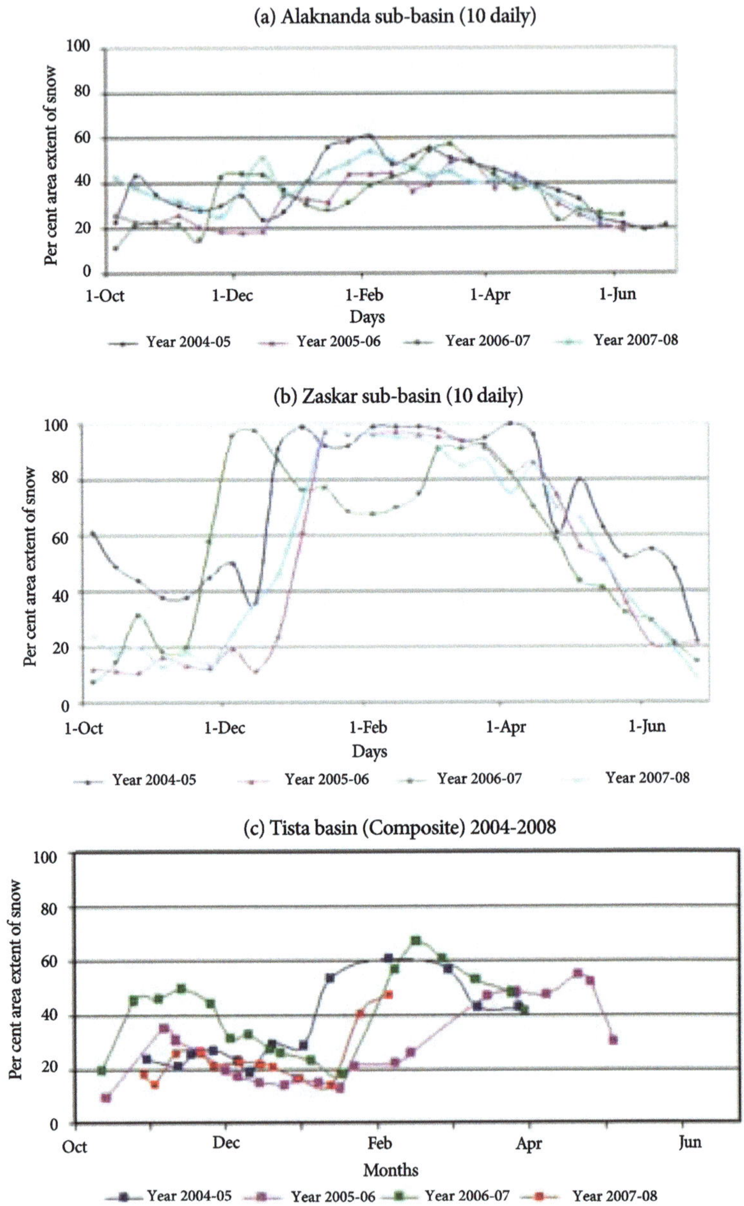

Fig. 12 Accumulation and ablation pattern of snow cover for the period 2004–08. *Source* Kasturirangan et al. (2011)

Figure 12a–c shows accumulation and ablation curves for three sub-basins namely, the Alaknanda (Ganga basin), the Zanskar (Indus basin) and the Tista (Brahmaputra basin) which are located in different climatic zones. The curves indicate that though the pattern of accumulation and ablation in different years are not very different in each basin but it differs from one basin to the other. This information is highly useful to estimate snow melt runoff in each basin. The pattern of Tista basin is influenced by south-eastern monsoon.

Detailed monitoring of snow cover pattern has also been carried out for Alaknanda, Bhagirathi and Yamuna sub-basins of Ganga from the year 2004 to 2012 covering the months October to June (Rathore et al. 2015). The snow products were generated using Normalized Difference Snow Index (NDSI) at a spatial resolution of 56 m. NDSI uses green and SWIR channel of AWiFS sensor. Minimum and maximum snow cover was found to be 998, 669, 141 km^2, and 7874, 5876, 3068 km^2 for Alaknanda, Bhagirathi and Yamuna sub-basins respectively. The areal extent of snow was found to be higher than the mean during the years 2004–05, 2007–08 and 2011–12 for all sub-basins. Mean of monthly fluctuations in between the maximum and minimum snow cover were recorded as 3105, 2305, 1235 km^2 corresponding to variation in snow line altitude of 1613, 1770, 1440 m respectively. A subtle increase in the snow cover has been observed in these three sub-basins during 2004–2012.

7 Conclusions and Future Scope

Inventory and monitoring of the Himalayan snow cover and glaciers have been carried out using satellite data. Inventory of the glaciers of the Himalayas (Indus, Ganga and Brahmaputra basins) has been done on 1:50,000 scale using Resourcesat-1 satellite data of 2004–07 time period. The total number of glaciers in these three basins put together is 32,392 covering an area of 71,182.08 km^2. About 60% of the glaciated area in the ablation zone is covered with debris which makes these glaciers protected from faster melting due to solar radiation. More than two thousand glaciers, well distributed in different climatic zones of the Himalaya, have been monitored for the period 1989–90 to 2001–04 and 2001–02 to 2010–11. While 76% of the glaciers have shown retreat, 7% have advanced and 17% have shown no change during the period 1989–90 to 2001–2004. Whereas only 12.3% glaciers have shown retreat, 86.6% of glaciers have shown stable front and 0.9 and have shown advancement during the next decade (2001–02 to 2010–11). Average retreat during the period 1989–90 to 2001–04 has been 3.75% in area, while the average retreat was only 0.2% in area during 2001–02 to 2010–11. This clearly indicates that the rate of retreat of glaciers in the Himalaya has come down during the decade 2001–02 to 2010–11 as compared to the previous decade (1989–90 to 2001–04). Mass balance of more than 700 glaciers of the Himalaya have been computed and monitored based on AAR approach. Large number of glaciers had negative mass balance during the years 2005, 2006 and 2007. However, the situation has changed and more than 50% of the glaciers have shown positive mass balance during the years 2010, 2011 and 2012.

Seasonal snow cover has been monitored for the Indus, Ganga and the Brahmaputra basins for the period 2000–2015. The Indus basin has shown an increasing trend in the snow cover for the month of September (end of the ablation) during the above period. However the other two basins did not show any trend in the snow cover.

Monitoring of Himalayan glaciers needs to be continued as they are important: (i) source of fresh water for human consumption and irrigation; (ii) source of energy (hydro power), (iii) regulate the climate and (iv) are critical indicators of climate change.

Apart from the inventory and monitoring of snow and glaciers of the Himalaya, there is a need to develop techniques/model to estimate the thickness of glaciers and snow using space data. Data from space borne Lidar and interferometric SAR need to be explored for the estimation of thickness of glacier ice and mass balance. Development of model for ice thickness estimation will require systematic field experimentation. Modelling energy balance over the glacier surface and its relationship with the melting/sublimation of ice is another important area of work which needs to be pursued.

Acknowledgements Author is thankful to Shri Tapan Misra, Director, Space Applications Centre (SAC), Ahmedabad for his keen interest in the work presented here. Discussions and help by Dr. I. M. Bahuguna, Mr. A.K. Sharma, Mr. S.K. Singh and Mr. B.P. Rathore, scientists at SAC is thankfully acknowledged. The author is also thankful to Dr. A.V. Kulkarni, Dr. Shakil Romshoo and Dr. Milap Sharma for fruitful discussions.

References

Ajai et al (2011) Snow and Glaciers of the Himalayas. Space Applications Centre, (ISRO), Ahmedabad, India. Also available on www.moef.nic.in

Bolch T, Menounos B, Wheate R (2010) Lansat based inventory of glaciers in western Canada, 1985–2005. Remote Sens Environ 114:127–137

Bahuguna IM, Rathore BP, Mohan A, Kulkarniand AV, Ajai (2010) Glacier cover change: a critical parameter for climate change studies. NNRMS Bull, NNRMS (B)-35, pp 1–179

Bahuguna I.M, Rathore BP, Brahmbhatt R, Sharma M, Dhar S, Randhawa SS, Kumar K, Romshoo S, Shah RD, Ganjoo RK, Ajai (2014) Are the Himalayan glaciers retreating? Current Science 106(7):1008–1013

Brahmbhatt R, Bahuguna IM, Rathore BP, Kulkarni AV, Nainwal HC, Shah RD, Ajai (2012) A comparative study of deglaciation in two contrasting basins (warwan and bhut) of western Himalaya. Curr Sci 163(3):298–304

Benn DI, Evans DJA (2010) Glaciers and Glaciation, 2nd edn. Hodder Education, London, p 795

Chaohai L, Sharma CK (1988) Report on first expedition to glaciers in the Pumqu (Arun) and Poiqu (Bhote—Sun Kosi) River Basins, Xi-Zang (Tibet) China. Science Press, Beijing, China, p 192

Dozier J (1984) Snow reflectance from Landsat-4 Thematic Mapper IEE. Trans Geosci Remote Sens 22(3):328–328

Dozier J (1989) Spectral signature of alpine snow cover from the landsat thematic mapper data. Remote Sens Environ 28:9–22
Flint RF (1964) Glaciers and quaternary geology. Wiley, New York
Hall DK, Chang ATC, Siddalingaiah H (1988) Reflectance of glaciers as calculated using landsat—thematic mapper data. Remote Sens Environ 25:311–321
Hewitt Kenneth (2014) Glaciers of the Karakoram Himalaya. Springer
Kaul MK, Puri VMK (1999) Inventory of the Himalayan glaciers: a contribution to the international hydrological programme. Geological Survey of India, p 165
Kasturirangan K, Navalgund RP, Ajai (2011) Observed changes in the Himalayan—Tibetian glaciers. Fate of Mountain in the Anthropocene. Pontifical Academy of Science, Vatican
Kulkarni AV, Singh SK, Mathur P, Mishra VD (2006) Algorithm to monitor snow cover using AWiFS data of Resourcesat for the Himalayan region. Int J Remote Sens 27(12):2449–2457
Kulkarni AV, Philip G, Thakur VC, Sood RK, Randhawa SS, Chandra Ram (1999) Glacial inventory of the Satluj basin using remote sensing technique. Himalayan Geol 20(2):45–52
Kulkarni AV (1992a) Glacier inventory in the Himalaya, natural resources management—a new perspective. National Natural resources management system. Bangalore, India, pp 474–478
Kulkarni AV (1992b) Mass balance of Himalayan glaciers using AAR and ELA methods. J Glaciol 38(128):101–104
Ostrem G (1975) ERTS data in glaciology—an effort to monitor glacier mass balance from satellite imagery. J Glaciol 15(73):403–415
Rathore BP, Singh SK, Bahuguna IM, Brahmbhatt R, Rajawat AS, Thapliyal A, Panwar A, Ajai (2015) A Spatio-temporal variability of snow cover in Alaknanda, Bhagirathi and Yamuna sub-basins Uttarakhand Himalaya. Current Science 108(7):1375–1380
Rastogi G, Ajai (2013a) Energy balance modeling for ablation estimation of Gangotri glacier. J Geomatics 7(2):178–185
Rastogi G, Ajai (2013b) Comparison of the energy balance on Gangotri and Chhota Shigri Glaciers. Int Arch Photogrammetry, Remote Sens Spat Inf Sci 40(8):537–542
SAC (2015) http://sac.gov.in/SACSITE/SAC-Flyers/menu-links/land/2.3
Sangewar CV, Shukla SP (2009) (Eds) Inventory of the Himalayan glaciers: a contribution to the international hydrological programme. Geological Survey of India, Special publication number 34
Sharma AK, Singh S, Kulkarni AV (2008) Approach for Himalayan glacier inventory using remote sensing and GIS techniques. In: Venkataraman G, Nagarajan R (eds) Proceedings of international workshop on snow, ice, glacier and avalanches. IIT-B. Tata McGraw-Hill Publishing Company Ltd, New Delhi, pp 177–185
Sharma AK, Singh SK, Kulkarni AV, Ajai (2013) Glacier inventory in Indus, Ganga and Brahmaputra basins of the Himalaya. Natl Acad Sci Lett 36(5):497–505
Singh SK, Rathore BP, Bahuguna IM, Ajai (2014) Snow Cover variability in the Himalayan-Tibetian region. Int J Climatol 34:446–452

Mitigation Strategies to Combat Climate Change in the Himalayan Mountains

M.R. Bhutiyani

Abstract The Himalayan Mountains have warmed significantly in the last century. The rate of warming during winter season is higher than in the monsoon. Consequent depletion of winter snow cover during winter and higher rates of summer melting of glaciers due to rising temperatures are likely to affect hydro-meteorological regimes of many river basins in Himalayas and livelihood of millions of people inhabiting the Indo-Gangetic Plains of Northern India. This paper explores the reasons behind climate change in Himalayas and addresses the issue of mitigation of these impacts by suggesting use of innovative latest state-of-the-art green technologies to tap renewable sources of energy such as solar, wind, hydro-electric and geothermal energy.

Keywords Climate change · Hydrological regimes · Mitigation measures

1 Introduction

The high mountain areas such as the Alps, the Rockies, and the Himalayas etc. are considered as the 'hotspots' over the surface of the earth where impacts of climate change are likely to be felt significantly. With regard to the Himalayas, their vulnerable ecosystem appears to have reacted to even the slightest possible changes in the temperature and precipitation conditions. Various studies on the climate change in the Himalayas have addressed some crucial issues such as, how sensitive are the Himalayas to the spectre of climate change now sweeping the globe by way of rising temperatures and changed precipitation regimes, the causes thereof and future manifestations in terms of changes in hydrological regimes of various Himalayan river basins. The cascading effects of these changes on the vast expanse of water existing in the form of glacier-ice and snow in Himalayas, the forest cover, the health and the socio-economic conditions of large populace inhabiting the

M.R. Bhutiyani (✉)
Defence Terrain Research Laboratory, Delhi, India
e-mail: mahendra_bhutiyani@yahoo.co.in

P.S. Goel et al. (eds.), *Science and Geopolitics of The White World*,
DOI 10.1007/978-3-319-57765-4_9

Indo-Gangetic plains, have been the issues of serious concern The studies have established that the Himalayan region has 'warmed' significantly during the last century at a rate, which is disturbingly higher than the global average (Yadav et al. 2004; Bhutiyani et al. 2007, 2010; Shekhar et al. 2010; Dimri and Dash 2012). Effects of climate change are evident on decreasing winter snowfall component in total winter precipitation on the windward side and some portions of the leeward side of the Pirpanjal Range since 1991, leading to delay in onset of winter and early spring and effective reduction in snowfall duration period. Coupled with rising summer temperatures, the glaciers are affected and they are in recession mode. This has given rise to a change in the nature of natural hazards in the area. Earth-flows and mud-flows, which were hitherto uncommon occurrences in the area, have been observed to occur in high altitude regions (>4500 m) over the glaciers (Bhutiyani et al. 2015). Consequently, these factors in combination have altered the discharge patterns in the Himalayan Rivers leading to an overall decreasing discharge in majority of Himalayan Rivers. These changes are attributed to enhanced anthropogenic (human related) activities in this region.

This paper describes the magnitude of warming in the Himalaya in the last century, reasons thereof and the role played by the greenhouse gases produced by the human-related activities. An attempt has also been made to explore the options before India, by way of unilateral mitigation strategies without affecting its development and poverty alleviation programmes and voluntary, if not legally binding, commitments to salvage the situation.

2 Magnitude of Climate Change in NW Himalayas

The analyses of the temperature data collected manually at different observatories during the period from 1866 to 2012 show significant rate of warming during the winter season (1.4 °C/100 years) than the monsoon temperature (0.6 °C/100 years). There is rapid increase in both, the maximum as well as minimum temperatures, with the maximum increasing much more rapidly. Annual rate of warming (1.1 °C/100 years) is abnormally higher than the global rate (about 0.7 °C/100 years) during this period. Not all regions of the NW Himalayas have reacted uniformly to the spectre of climate change. Studies have confirmed significant spatial and temporal variations in magnitude of winter as well as summer warming in different ranges. While windward side of the Pirpanjal and parts of Greater Himalayan and Karakoram ranges have shown statistically significant winter and summer warming, leeward sides of these ranges have not shown much change. The most remarkable finding of this study is the significant decreasing trend experienced at almost all stations above equilibrium line (>5300 m in altitude) in winter warming as well as winter precipitation in higher reaches of the Karakoram Himalayas in last three decades (Bhutiyani 2016).

From the precipitation point of view, significant decreasing trends (at 95 per cent confidence level) in the monsoon and overall annual precipitation during the study period are indicated. In contrast, the winter precipitation has shown an increasing but statistically insignificant trend (at 95% confidence level). Rising winter air temperatures have caused decreasing snowfall component and increasing rainfall component in total winter precipitation on the windward side of the Pirpanjal Range, parts of Greater Himalayas and the Karakoram ranges. The analyses also show that although winter precipitation in the Himalayas has remained trendless in last 140 years, there are significant increasing trends in the extreme snowfall events during winters and rainfall events during summers in Pirpanjal and Shamshawari Ranges in last three decades and insignificant but increasing trends in the Great Himalayan and Karakoram Range (Shekhar et al. 2010; Dimri and Dash 2012). Decrease in winter snowfall amounts and increasing rainfall component at almost all stations has been affected to some extent, by the increase in winter air temperature during this period.

The spatial and temporal variations in winter and summer warming and consequent precipitation changes in different ranges/regions of the NW Himalayas are attributed to varying scales of anthropogenic activities and growing urbanization of the areas. Decreasing temperatures in last three decades in the Karakoram Himalayas with altitudes above the equilibrium line (>5300 m) are attributed to prevalence of permanent snow cover which appears to have influenced their micro-climatology. These studies have significant bearing on the mass balance of the glaciers in the region and the hydrological behaviour of various river systems in the Himalayas (Bhutiyani et al. 2008).

3 Strategies of Mitigation of Climate Change in NW Himalayas

It is evident from the foregoing discussion that depletion of snow cover during winters and shrinkage of glaciers during summers have critically affected the hydrological regime/discharge patterns of rivers in the Himalayas. Such changes pose a serious challenge to the developing country like India which experiences significant climatic variability affecting large agrarian population dependent on vicissitudes of monsoonal precipitation in plains and mountainous regions. Fortunately, the Himalayan Mountains have not reached the tipping point yet, but are already showing signs of weariness. This may be attributed to multiple reasons such as comparatively lower population densities and modest rates of urbanization. In the light of increasing environmental degradation in last few decades, focused efforts are required to identify the root causes which may have led these fragile mountains to the brink of catastrophe and avoid the consequent disastrous after-effects. Significant remedial and preventive measures at technology, socio-economic, community and policy level may have to be implemented to first, salvage the situation and then arrest the warming trend. It is now essential to formulate certain mitigation strategies to combat climate change in NW Himalaya which focus on monitoring and subsequent reduction in greenhouse concentration in these valleys. Such strategies should address some important issues like:

- How to combat winter and overall warming and restore the health of our receding glaciers.
- How to arrest the overall environmental degradation taking place because of uncontrolled urbanization and consequent increasing concentration of greenhouse gases in the atmosphere.
- How to adopt green technologies.

While it may not be possible to completely reverse the trends in level of urbanization and anthropogenic activities because of rising local population, heavy influx of tourists and consequent vehicular traffic and production of greenhouse gases and their effects, current trends and 'business-as-usual' approach may spell disaster for the fragile ecology of the Himalayan Mountains in the long run. With some conscious and sincere efforts, it may be possible to strike a healthy balance between the preservation of Himalayan ecosystem and developmental needs of the local population, meaningful exploitation of resources and poverty alleviation.

Studies have shown that increasing winter temperatures is the most important factor in overall warming process of the Himalayan Mountains. It is extremely essential that unique strategies are devised to decelerate abnormally high rate of winter warming and save depleting winter snow-cover. Presence of a blanket of snow-cover on the ground alters the net energy balance and micro-climatology of the basins considerably. Higher values of reflectance of freshly fallen snow (in comparison to exposed ground) with lower surface temperatures ensure substantially lower contribution of net shortwave radiation ('insolation') and reduced losses due to lesser outgoing long-wave 'terrestrial' radiations to the net energy balance. These lead consequently to lower ambient air temperatures in the adjoining atmosphere. With the onset of spring, the snowline recedes at a faster pace leaving behind exposed ground with lower reflectance values, rate of increase in air temperature has been observed to be much higher, This effect has been termed as 'Mountain Spring Heat Island effect' (Bhutiyani 2016). Although precipitation regime of an area with its temporal variability and snowfall and rainfall components is a natural phenomenon and it cannot be controlled, the duration of snow cover on ground can partially be governed by extraneous anthropogenic factors such as changing the land use pattern, diminishing degree of urbanization etc. To address this issue, there is a need to adopt a multi-fold strategy which includes:

- Identification of vulnerable river basins/areas: Identification of such river basins where 'Mountain Spring Heat Island effect' can have dangerous effects on their ecological and hydro-meteorological regimes would be an important first step in this direction. Such basins will have to be selected based on their magnitude of warming, degree of urbanization, concentration of glaciers and anthropogenic activities in them. River basins on windward and leeward side of the Pirpanjal and Shamshabari Ranges like, the Beas, Ravi, Satluj and Chenab can be taken up on 'top-priority basis'.

- Prolonging duration of winter snow-cover: Snow harvesting at multiple sites can be resorted to by simple techniques of snowfall pits and trenches, snow-rakes, and snow catchment dams (Fig. 1). These can serve a dual purpose of acting, both as avalanche control and snow storage structures to elongate effective duration of winter snow cover. A remarkable attempt by Mr. Chawang Norphelat snow-harvesting and creation of temporary artificial glacier and 'ice stupas' in a north-facing rivulet at Nang near Leh (Fig. 2) can be replicated at many places where similar conducive topographic and meteorological conditions exist.
- Adopting techniques of artificial reduction in surface reflectivity: Large-scale urbanization and deforestation, particularly in the Siwalik, Pirpanjal and parts of greater Himalayas, have altered the hilly landscapes substantially. Besides removing healthy vegetation cover that adorned these mountain ranges, they are now covered at places with either greyish concrete roof-tops of house-holds or have exposed the barren ground beneath with very low surface reflectivity or albedo values (Fig. 3). This modification has resulted in drastic change in radiation regimes of such areas and has contributed immensely in the warming process. Suitable techniques are needed to be devised and adopted as a mitigation strategy at a policy level to induce "negative radiative forcing". Time-tested method like covering the roof-tops with highly reflecting material or painting them white in all river basins which are sensitive to climate change and where the chances of their micro-climatology being affected adversely are very high, can be an effective mitigation strategy (Fig. 4).

Fig. 1 Avalanche control structures for controlling avalanches and harvesting the Snow. *Source* Author

Fig. 2 Creation of artificial glacier at nang near leh and 'snow-stupa' as an ice and snow harvesting method. *Source* Author

Fig. 3 Highly urbanized and deforested barren ridge-tops in the siwalik, pirpanjal and Greater Himalayan Ranges. *Source* Author

This method has been strongly recommended by Dr. Steven Chu, the former US Secretary of Energy and a Nobel prize-winning scientist. If adopted in true spirit, it would be the equivalent of taking all the cars in the world off the road for 11 years. Retrofitting urban roofs and pavements in tropical and temperate regions such as the

Note: White-painted roof-tops of residential buildings can be one of the best strategies to fight climate change in Himalayas. Source: www.thinkprogress.org/climate

Fig. 4 White-painted roof-tops of residential buildings as best statergy to fight climate change in Himalayas. *Source* Author

Himalayas with solar reflective materials would offset about 44 billion tons of carbon dioxide. It is also expected to lower the cost of air conditioning, making buildings more comfortable and mitigate the 'Mountain Spring Heat Island effect' caused by the concentration of concrete surfaces in cities in Himalayan Mountains. A mass movement on similar lines of 'white-roof' movement in United States needs to be started in Himalayan states to mitigate effects of climate change.

- Managing wastage heat in fragile high-altitude areas: Significant amount of wastage heat is produced by the uncontrolled heavy vehicular traffic and a large number of cook-houses (dhabas) (Fig. 5) at high-altitude locations which attract large number of tourists every year. These also generate dangerous greenhouse gases and black carbon/soot particles which alter the energy balance and cause warming. Rotting wastage food and large amount of human and animal waste lead to production of methane, a potent greenhouse gas in large quantities. While it may not be possible to altogether stop the vehicular traffic and remove these eateries from the Himalaya, resorting to controlled tourism and vehicular traffic, provision of hybrid/electric/CNG vehicles, removal and relocation of cook houses/dhabas from highly vulnerable areas to other convenient areas, will go a long way in arresting warming and downward spiral of environmental degradation in Himalayan Mountains (State of the World 2009). Such mitigation methods adopted in eco-sensitive areas such as Rohthang Pass, Marhi, Dhundi, Solang Nala in Beas basin and similar areas in other river basins and Kashmir Valley have yielded partial but positive results. Concerted and conscious efforts would be required to find out area-specific solutions to the problems. At places

Fig. 5 Heavy vehicular traffic leading to the rohthang pass (Himachal pradesh) and unplanned cook-houses and eateries as significant sources of greenhouse gases concentration. *Source* Author

such as Rohthang Pass and other high altitude locations visited by large number of tourists, wherever feasible, options like provision of ropeways and electric cable cars such as at Gulmerg in Kashmir Valley and at Solang in Himachal Pradesh should work wonders for the frail ecology of the area.

- Use of Renewable 'Low-Carbon' sources of energy: In the Himalayan perspective, there is an urgent need to look at climate change as an opportunity, rather than a 'liability'. Development of latest state-of-the-art greener technologies would give a quantum impetus to the economy and consequently the GDP of the country. Adopting these technologies shall enable a smooth shift from fossil-fuel based 'high-carbon' economy to a 'low-carbon' economy. With multiple options to choose from, extensive use of renewable sources of energy such as solar, hydel power, wind and geothermal energy (Fig. 6) may lead to a 'low-carbon' environment and drastic reduction in greenhouse emissions. To meet the requirements of area specific end-use applications, integration of two or more renewable energy sources/generation systems may be thought of as a solution.

Exploitation of geothermal energy available for generation of electricity and for space heating/cooling in rugged Himalayan terrain offers a promising alternative to arrest climate change. The technology to tap of this unlimited amount of geothermal

Fig. 6 Renewable sources of energy as good option in Himalayan mountains to combat climate change. *Source* Author

energy is already available in the Indian market. A pilot plant involving installation of geothermal heat pump (GHP) and ground source heat exchanger to make use of this source for space heating has been functional at HQ SASE, Manali. A similar plant with additional capacity of electric power generation is operational at Puga in Leh valley in Ladakh. Further research needs to be carried out to improve the efficiency of these systems and make them economically viable for large scale utilization.

In addition to these, intensive research efforts will have to be devoted towards development of alternative fuels like hydrogen cells to run future vehicles.

- Carbon capture and sequestration (CCS): This particular method has not been tried so far in India, but has an incredible potential to play an important role in mitigation of impacts of climate change in the Himalayan Mountains. This technology can capture up to 90 per cent of the carbon dioxide (CO_2) emissions produced from the use of fossil fuels in electricity generation and industrial processes, preventing the carbon dioxide from entering the atmosphere (Kunzig and Broecker 2009).The carbon dioxide is then safely stored in deep underground geological rock formations in the form of suitable carbonates. Use of CCS with renewable biomass is one of the few carbon abatement technologies that can be used in a 'carbon-negative' mode—actually taking carbon dioxide out of the atmosphere (Fig. 7). To achieve this objective, focused research efforts, specific to the Himalayan environment, may be required to be carried out to develop contraptions called as 'CO_2 scrubbers' which can be deployed at high altitude vulnerable regions.

Afforestation. Forests are the largest storehouse of carbon, after coal and oil and they also act as one of the most effective 'carbon-sinks' of the nature after the oceans. When they are destroyed–by logging and clearing for developing–massive

Fig. 7 'Carbon scrubber' at vulnerable locations in the Himalayas to crub carbon dioxide. *Source* www.canadianmanufacturing.com

quantities of CO_2 is released into the atmosphere. Large-scale afforestation of the presently barren stretches of Himalayan Mountains would go a long way in restoring the balance and arrest climate change in Himalayas.

4 Conclusion

There is perceptible evidence of climatic change with warming trend in temperature in the Himalayan Mountains. Significant warming has been observed during winters. Consequent depletion of winter snow-cover and glaciers due to rising temperature may affect the hydro-meteorological regimes of river basins and livelihood of millions of people inhabiting the Indo-Gangetic Plains of Northern India. Urgent, unilateral, proactive and constructive methods are needed to mitigate the effects of climate change. Urgency and vision on part of politicians, planners and policy makers, science and technology community are the two pillars on which the humanity's hope now hangs to salvage the situation.

References

Bhutiyani MR, Kale VS, Pawar NJ (2007) Long-term trends in maximum, minimum and mean annual air temperatures across the Northwestern Himalaya during the 20th century. Clim Change 85:159–177

Bhutiyani MR, Kale VS, Pawar NJ (2008) Changing stream flow patterns in the rivers of the Northwestern Himalaya: implications of 'global warming' in the 20th century. Curr Sci 95 (5):618–626

Bhutiyani MR, Kale VS, Pawar NJ (2010) Climate change and the precipitation variations in the Northwestern Himalaya: 1866–2006. Int J Climatol 30(4):535–548

Bhutiyani MR, Ravikant, Rajinder Prashad (2015) Mass movement events in upper karakoram Himalayas: implications of climate change. J Eng Geol 40(1):7–19

Bhutiyani MR (2016) (forthcoming) Spatial and temporal variability of climate change in high-altitude regions of NW Himalaya. In: Singh RB (ed) Climate change and dynamics of glaciers and vegetation in the Himalaya: towards future earth initiatives. Springer

Dimri AP, Dash SK (2012) Wintertime climatic trends in the Western Himalayas. Clim Change 111:775–800

Kunzig R, Broecker W (2009) Fixing climate. Green profile in association with sort of books

Shekhar M, Chand H, Kumar S, Srinivasan K, Ganju A (2010) Climate-change studies in the Western Himalaya. Ann Glaciol 51:105–112

State of the World (2009) The World Watch Institute, Indian Edition. Center for Environment Education Washington, USA.

Yadav RR, Park WK, Singh J, Dubey B (2004) Do the Western Himalaya defy global warming? Geophys Res Lett 31:L17201

www.canadianmanufacturing.com

www.thinkprogress.org/climate/2009/05/27/204160/energy-steven-chu-white-roofs-geo-engineering-adaptation-mitigation/

Winter Climate and Snow Cover Variability Over North-West Himalaya

H.S. Negi, M.S. Shekhar, H.S. Gusain and A. Ganju

Abstract The impact of climate change over Indian Himalaya has received a great deal of attention worldwide. The present study focuses the recent winter climate, snow cover and albedo variability over north-west Himalaya (NWH). We analyze the trends of climate and snow cover variations in past two decades (1991–2011) and compare with the recent years 2001–2014, in three different snow climatic zones of NWH viz. Lower Himalaya (LH), Great Himalaya (GH) and Karakoram Himalaya (KH). The analysis of past two decades suggests that the maximum temperatures have increased and minimum temperatures have decreased in all the three zones of NWH. The trends of winter mean temperatures have been found increasing in LH and GH, and decreasing in KH. The increasing trend in diurnal temperature range (DTR) was observed over all the three zones and attributed to increasing maximum and decreasing minimum temperatures. Significant increasing trends in total precipitation (solid and liquid) were observed in LH and GH, and insignificant decreasing trend in KH. The field observed winter snow albedo of the same period suggest significant decreasing trend in GH and no trend in LH. The snow cover area (SCA) and albedo was monitored using MODIS sensor for the winter period 2001–14. These recent years (2001 onward) analysis suggests that the trends of SCA and snow albedo are insignificantly increasing over NWH. This has been supported by insignificant decreasing winter mean temperatures in all the three zones or vice-versa. These decreasing trend in temperature and increasing trends in SCA and albedo are attributed to recent 'Hiatus' in rising winter mean temperatures and thus global warming. The trends of these climatic variations in different periods also support the climate change impact on Himalayan snow cover and glaciers.

Keywords Himalayas · Albedo · Snow cover · Climate change

H.S. Negi (✉) · M.S. Shekhar · H.S. Gusain · A. Ganju
Snow and Avalanche Study Establishment, Him Parisar, Sector-37A, 160036, Chandigarh, India
e-mail: hs.negi@sase.drdo.in

P.S. Goel et al. (eds.), *Science and Geopolitics of The White World*,
DOI 10.1007/978-3-319-57765-4_10

1 Introduction

Recent climate change and global warming over different parts of the world has raised concern about the snow cover and glaciers reduction with alarming rate. Regional scale forcing induced by the regional topographical features along with the land use characteristics modifies the global effect of the climate change. The changing of climate and warming of the atmosphere has a large impact on the water resources, agriculture and overall economy of the country. Thus it becomes necessary to study different aspects of the climate change based on the available data in long term period over the region. Several observational studies show significant temperature and precipitation changes over India in long term period. A recent study shows that all India mean annual temperature has increased by 0.5 °C during the period 1901–2003 (Kothawale and Rupa kumar 2005). This increase in temperature is consistent with the global warming over the country. The trend in the pre-monsoon (March–May) temperature over the western Himalaya has been studied by Yadav et al. (2004). From the observational and reconstructions from the tree rings, they found that the pre-monsoon minimum temperature shows decreasing trend during the later part of the twentieth century. They also found that the rate of decrease of minimum temperature is three times than that of maximum temperature thereby responsible for the cooling trend in the pre-monsoon mean temperature. The temperature and precipitation trends over North and South India for different phases of monsoon were carried out by Dash and Hunt (2007). They found large differences in trends in minimum temperature and cloud amounts between North and South India and asymmetry in the increasing temperature trends between different seasons in a year. Dash et al. (2007) also found reduction in the summer monsoon (June–September) rainfall over India and suggested that reduction of summer monsoon rainfall in the warming atmosphere can be explained by both climate change and mesoscale effects of the mountains on the monsoon flow.

The snow and glaciers in Himalayas are considered as an important source of fresh water. Therefore, it is very important to study the impact of climate change on the snow cover and glaciers of these regions. The NWH receives precipitation in the form of snow during winter (November–April) due to the passage of synoptic system known as Western Disturbance (WD) (Pisharoty and Desai 1956). In the complex mountainous regions of the NWH, a limited number of studies have been carried out to analyze the temperature and precipitation trends in the global climate change scenario. Long term trends in the maximum, minimum and mean temperatures over northwest Himalaya during twentieth century carried out by Bhutiyani et al. (2007) suggests significant rise in the air temperature in the northwest Himalaya with winter warming at a faster rate. The study also reveals that real warming started from late 1960s and highest rate of increase was found in the last two decades of twentieth century. The winter time simulation of temperature and precipitation over the western Himalaya region has been studied by Dimri and

Ganju (2007). They simulated temperature and precipitation by using a regional climate model and found that the temperature represents cold bias and the precipitation is overestimated over the Himalayan region. However, studies by Fowler and Archer (2006) shows that there is a decreasing trend in mean annual temperature over the Upper Indus Basin (UIB) of the Himalaya since 1960s with more strongly since 1970s.

The Diurnal Temperature Range (DTR), which is the difference in daily maximum and minimum temperature, provides useful information for studying climate variability and climate change (Braganza et al. 2004). Studies indicate that in recent years, a vast land area of the world has recorded decreasing trend in DTR due to increase in minimum temperature in comparison to the maximum temperature (Easterling et al. 1997; Vose et al. 2005). Although a large number of studies have been carried out to study the relationship between DTR and various parameters over different parts of the world, very few studies have been carried out related to DTR over the Indian Himalayan region (Yadav et al. 2004; Roy and Balling 2005; Jhajharia and Singh 2011). This may be due to the fact that Himalayan region is a data sparse region. Increasing DTR has been observed in all the seasons over the Karakoram and Hindu Kush Mountains of the Upper Indus basin during the period 1961–2000. Scholars have observed increasing trends in the DTR due to a higher rate of increase in maximum temperature than minimum across the NWH region over the majority of river basins of northwest and central India (Fowler and Archer 2006; Bhutiyani et al. 2007; Singh et al. 2008).

Numerous studies have been carried out for mapping and monitoring of snow cover area (SCA) over Himalayas using satellite data (Immerzeel et al. 2009; Negi et al. 2009; Gurung et al. 2011; Singh et al. 2014). The presence of snow cover extends in high elevation regions and its interaction with the atmosphere can affect surface air temperature. Therefore snow-albedo feedback is a contributing factor for amplifying warming over Cryosphere region. Thus, it is important to understand the reasons for such spatial variability and the potential role of the snow-albedo feedback mechanism. In addition, in the recent past it has been reported that the Great Himalayan glaciers are receding and the Karakoram glaciers are showing the heterogeneity (Hewitt 2005; Bolch et al. 2012; Bhambri et al. 2013). Moreover the Himalayan glaciers are reported to be highly affected by the snow/ice darkening effect due to black carbon (BC) and dust, and these are attributed as significant factors for melting snow and glaciers (Qian et al. 2011; Ming et al. 2012). Ground-based observations of snow albedo are scarce in Himalaya mainly owing to the rugged terrain; this leads to a challenge in studying the regional variability in snow cover and albedo. In this paper, spatial and temporal variability of the trends of temperatures, total precipitation and snow albedo data from field observations over different zones of NWH have been analyzed. The SCA and albedo was monitored using satellite derived MODIS data of recent years 2001 onward. Further we discuss the impact of climate change on Himalayan Cryosphere in recent decades.

2 Data and Methodology

Snow and Avalanche Study Establishment (SASE) have a network of snow-meteorological observatories over NHW. Sharma and Ganju (2000) characterized the NWH on the basis of snow climatic conditions in three zones viz. lower, middle and upper Himalayan Snow Climatic Zones. The lower Himalayan zone is characterized by moderate temperature, high precipitation and significant changes in the snow characteristics occur here due to moderate temperature. The middle Himalayan zone is characterized by cold temperatures and mostly glaciated terrain. The upper Himalayan zone is extremely cold, receives dry snow and has large area covered by glaciers and permanent snow which are somewhat close to continental snow conditions. In the present study, we have adopted same classification and hereafter lower Himalayan, middle Himalayan and upper Himalayan zones are referred as Lower Himalaya (LH), Great Himalaya (GH) and Karakoram Himalaya (KH), respectively. A map of the study area with total 19 field stations used to collect different observations is shown in Fig. 1.

To study climate change during past two decades (1991–2011), the data of temperatures, snowfall and rainfall from 16 field stations spread over all the three zones are considered. The years from which consistent winter climate data (November to April) are present over all the stations have been considered for analysis. The snow

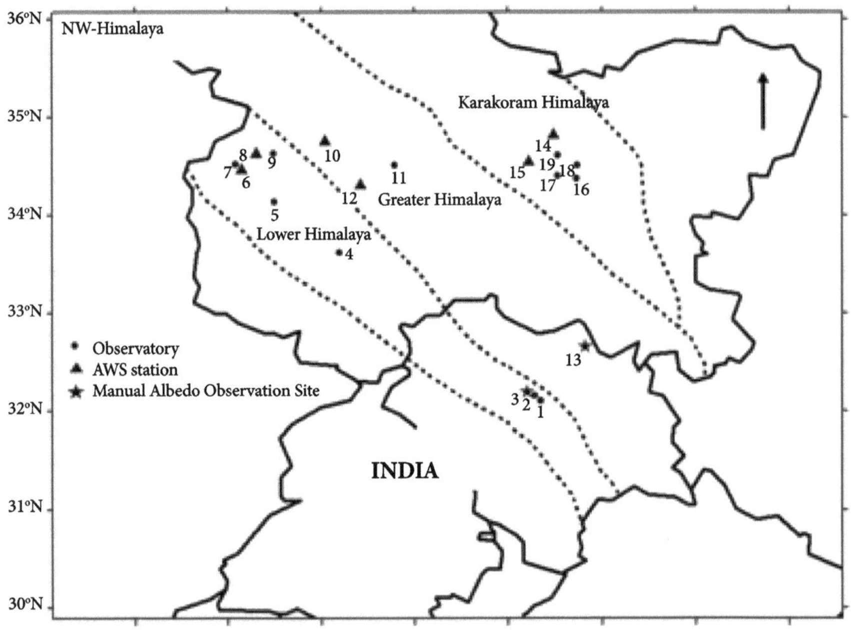

Fig. 1 A map of study area showing locations of field observation sites in different Himalayan zones of north-west Himalaya

albedo measurements were conducted manually (semi-automatic) at SASE field observatories at Patsio (3800 m) in GH and Dhundi (2900 m) in LH during January–April (or before if snowpack decayed early) (Fig. 1). It may be noted that the time period of the albedo data is slightly different from winter climate data. Albedo data of four winter months from January to April was used in the study as sufficient snow remains in all observatories during this period in all three NWH zones and can be compared. Another set of recent year's data between 2001 and 2014 from MODIS satellite data was used for the SCA and albedo analysis. The SCA maps of clear days were generated for all the three zones. The daily MODIS albedo products (MOD10A1, version-5), available from the National Snow and Ice Data Centre (http://nsidc.org/data/mod10a1.html) were analyzed of 08 locations (06 AWS and 02 Manual Observatories) (Fig. 1).The albedo products were obtained between 2000 and 2014 of 04 months (January and April), and analysed with a criteria of cloud free scenes by the MODIS cloud mask and as 100 per cent snow-covered by the MODIS snow algorithm (Klein and Stroeve 2002).

The daily mean temperatures have been calculated by the simple average of Tmax and Tmin and DTR have been calculated by taking difference of Tmax and Tmin. Albedo measurements were collected daily on an hourly basis between 0900 and 1400 h and daily mean albedo values were calculated. Cloudy days' albedo data were filtered manually by observing the cloud amount in morning (0830 h) and evening (1730 h) time and only clear days' albedo values were used. The linear trend analysis of the different parameters has been carried out by Mann–Kendall test with p-value to examine the significance of temporal variations.

3 Results and Discussions

3.1 Climate Observations in Last Two Decades (1991–2011)

3.1.1 Winter Temperatures

The seasonal average maximum temperature (Tmax) for the three different NWH zones is depicted in Fig. 2. The trends of Tmax for the all three zones were found to be positive. The slope and the level of significance are shown in the Table 1. The rise in Tmax for LH, GH and KH was found to be approx. 0.32, 1.4 and 1.0 °C respectively in 21 years (1991–2010). However, the increase in trends over these zones is not found significant. The spatial variability in Tmax over different zones of NWH can be well observed in Fig. 2.

The seasonal average minimum temperature (Tmin) for the NWH zones is shown in Fig. 3. All the zones showed decreasing trends of Tmin during winter. LH shows decrease in Tmin by ~0.8 °C, GH by ~1.3 °C and KH by ~1.4 °C (Fig. 3 and Table 1). The decreasing trends observed over these zones of NWH for Tmin have remained insignificant during the past two decades. Dash et al. (2007) found an increase of 0.98 °C in annual maximum temperature over the western Himalaya.

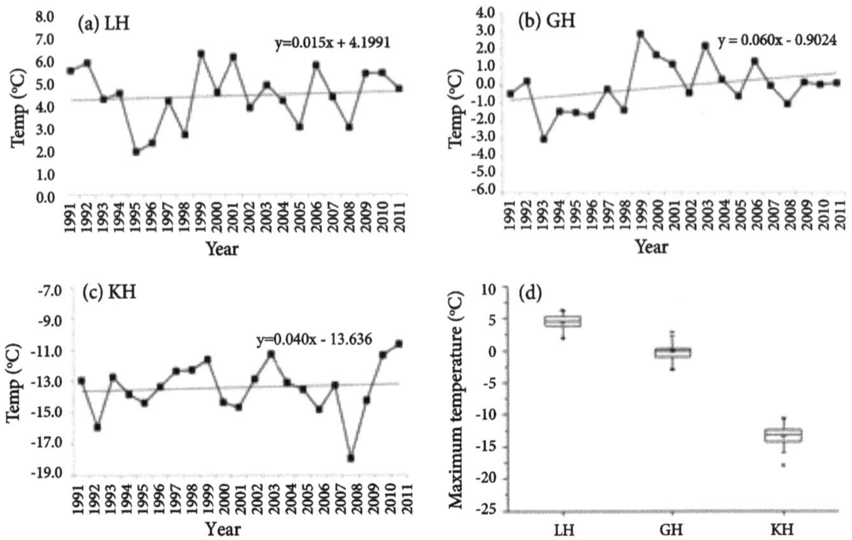

Fig. 2 Temporal variation in winter average maximum temperature in different zones of NWH **a** LH, **b** GH, **c** KH and **d** spatial variation of maximum temperature in different zones using box plot (*Data* SASE Observatories)

For annual minimum temperature, they found a decrease of ~1.98 °C during 1955–72, followed by an increasing trend over more recent decades. Shekhar et al. (2010) also reported increase in winter maximum temperature for LH (comprising Pir Panjal and Shamshawari ranges) and GH for a study period of 1988–2008, but they observed decreasing trend of Tmax for KH. Further they reported increase in the winter minimum temperature in LH, GH and decrease for KH.

The winter mean temperatures (Tmean) are depicted in Fig. 4.

LH and GH show increasing trends of Tmean during winter. The increase in Tmean over these regions is ~0.1 and ~0.6 °C respectively during the study period (Fig. 4 and Table 1). KH shows decreasing trend of Tmean (~0.9 °C) for the same duration. This may be due to the dominant contribution of Tmin compare to Tmax over the region. The seasonal average of DTR shows increasing trends in DTR for all the zones of NWH (Fig. 5 and Table 1). The increase in DTR for LH, GH and KH was found to be approx. 1.0, 3.5 and 3.2 °C respectively. Thus it can be inferred that the increasing trends of Tmax and decreasing trends of Tmin are contributing parameters for DTR increase and this is not due to the increase in Tmax only. DTR study has been carried out over the Indian subcontinent by Kumar et al. (1994). They showed that increases in DTR are due to the increase in Tmax, with Tmin remaining almost trendless. Yadav et al. (2004), showed that the cooling in pre-monsoon Tmean and warming in DTR is due to the decrease in Tmin and an increase in Tmax in the western Himalayas during later 20th century. Thus, DTR may be considered as an important parameter for climate change studies in comparison to Tmean.

Table 1 Test significance of linear trends using Mann-Kendall test for tmax, tmin, tmean, dtr, Precipitation and Albedo in different zones of NWH during 1991–2011

NWH Zone	Tmax		Tmin		Tmean		DTR		Precipitation		Albedo	
	Slope	*p*-value	Slope	*p*-value	Slope	*p*-value	Slope	*p*-value	Slope	*p*-value	Slope	*p*-value
LH	+0.015	0.789	−0.038	0.325	+0.005	0.929	+0.049	0.142	+3.108	0.009**	−3.8 E-04	0.894
GH	+0.069	0.178	−0.064	0.789	+0.028	0.613	+0.167	0.325	+1.494	0.049*	−0.009	0.001**
KH	+0.048	0.613	−0.069	0.178	-0.045	0.456	+0.153	0.111	−0.694	0.334	–	–

where
*Significant at $p < 0.05$
**Significant at $p < 0.01$

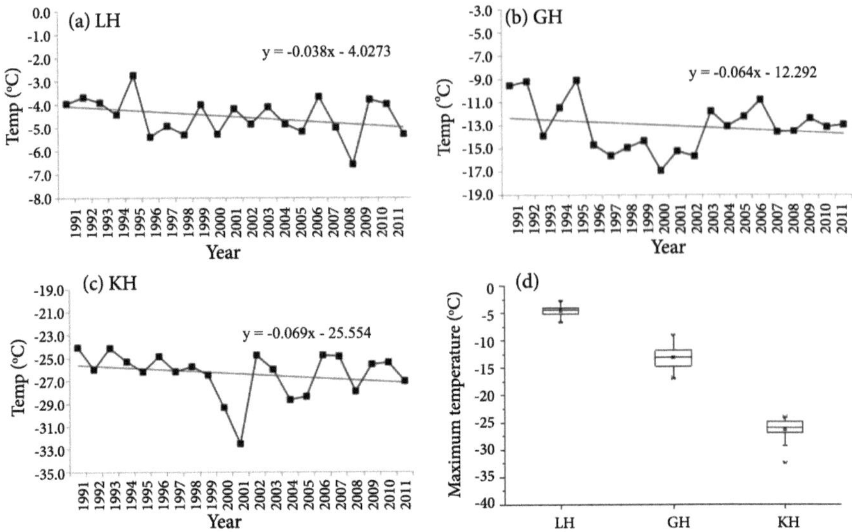

Fig. 3 Temporal variation in winter average minimum temperature in different zones of NWH **a** LH, **b** GH, **c** KH and **d** spatial variation of minimum temperature in different zones using box plot (*Data* SASE Observatories)

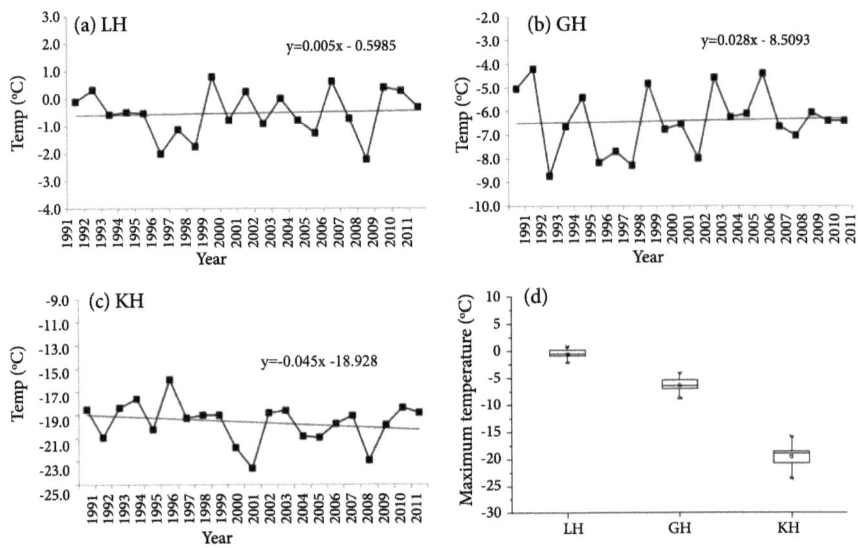

Fig. 4 Temporal variation in winter mean temperature in different zones of NWH **a** LH, **b** GH, **c** KH and **d** spatial variation of winter mean temperature in different zones using box plot (*Data* SASE Observatories)

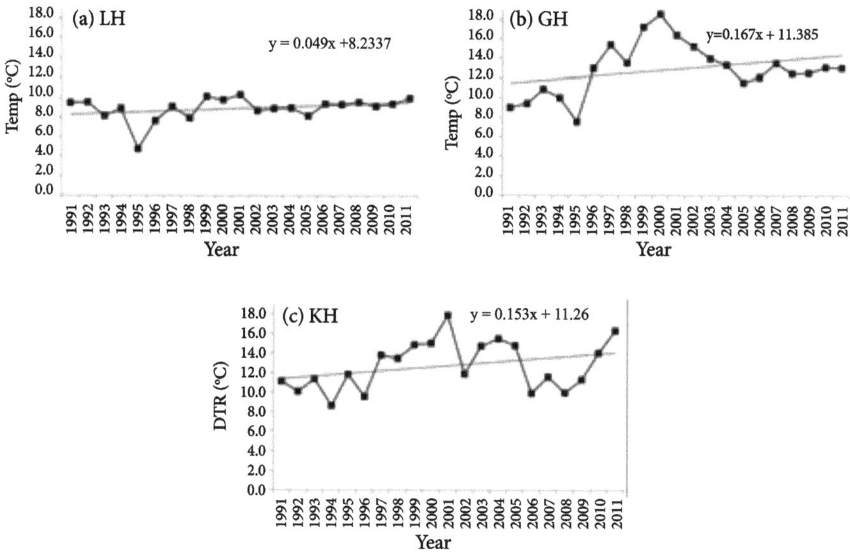

Fig. 5 Temporal variation in winter average DTR in different zones of NWH **a** LH, **b** GH, **c** KH (*Data* SASE Observatories)

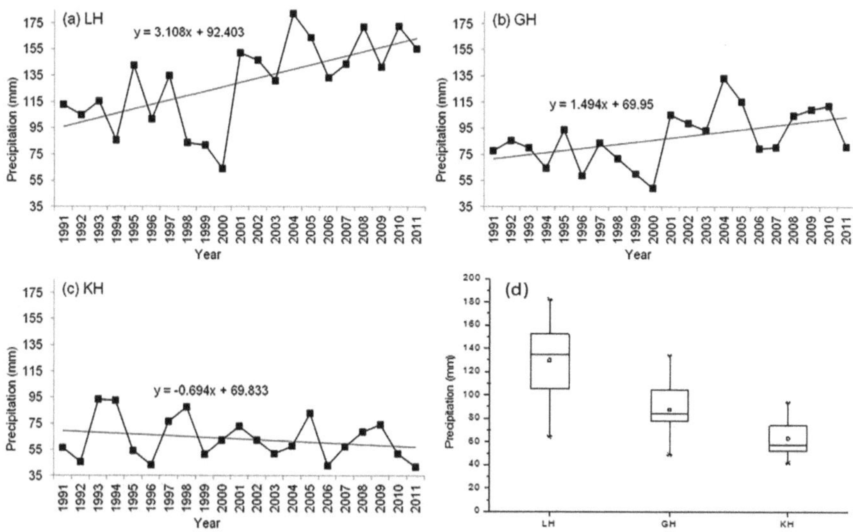

Fig. 6 Temporal variation in winter average precipitation in different zones of NWH **a** LH, **b** GH, **c** KH and **d** spatial variation of precipitation in different zones using box plot (*Data* SASE Observatories)

3.1.2 Winter Precipitation

The mean winter precipitation patterns indicate significant increasing pattern for LH and GH (Fig. 6 and Table 2). Whereas insignificant decreasing pattern of winter

Table 2 Test significance using Mann-Kendall test for SCA, Albedo and T mean in different zones of NWH during 2001–2014

NWH Zone	SCA		Albedo		Tmean	
	Slope	*p*-value	Slope	*p*-value	Slope	*p*-value
LH	+0.414	0.204	+1.7E-04	1.0	−0.090	0.351
GH	+0.225	0.578	+0.003	0.127	−0.127	0.139
LH	+0.633	0.175	+0.002	0.193	−0.129	0.250

precipitation was observed for KH. The monthly average winter precipitation with standard deviation (SD) values for LH, GH and KH was found to be 129 mm (SD:31 mm), 88 mm (SD:2 mm) and 63 mm (SD:16 mm) respectively. These observations quantitatively support the decrease in precipitation while traversing from LH to KH and the spatial variability in winter precipitation in different zones of NWH. The climatology of fresh snowfall over western Himalaya has been studied by Shekhar et al. (2010). Their study indicates decreasing trends in snowfall in all the ranges of western Himalaya. The decreasing trends in snowfall and increasing trends in total precipitation clearly indicates the increase in liquid precipitation i.e. rainfall in different ranges of Himalaya and this has been attributed to rise in mean temperatures in LH and GH. The trend in precipitation in KH remains decreasing and this may be due to the fact that this zone is situated in higher altitude having very low temperature and gets precipitation throughout the year in the form of snowfall only.

3.1.3 Winter Snow Albedo

A significant decrease in albedo of GH approximately of 0.2 was observed between 1993 and 2010 (Fig. 7a). The rise in winter mean temperature was found significant of approx. 1.6 °C in 18 years over GH. The temperature condition of GH (mean air temperature lies between −10 and 0 °C; Fig. 4d) with increasing trend was attributed to decreasing albedo due to thermal metamorphism, as the albedo reduction varies significantly for snow temperatures above −10 °C, where the metamorphic processes have profound effects on albedo (Aoki et al. 2003) form of snowfall only. No change was observed in albedo of LH in past two decades (1991–2010) (Fig. 7b). Though the rise in winter mean temperature in LH was found but as the snowfall in this zone occurs in warmer temperature regime, the Tmax was always found above 0 °C and Tmin was found subzero (Figs. 2d and 3d), which caused the melt-freeze metamorphism to proceed quickly and thus form clusters from snow single grains. After multiple melt-refreeze cycles, these clusters transform into single melt-freeze grains of much larger grain radius (Colbeck 1982). Because of frequent melt-refreeze cycles, snow cover of LH is characterized by coarse grain size distribution with moist to wet snow. Therefore snow albedo found to be lower in LH. Such moderate winter mean temperatures of LH station were found near 0 °C since beginning of the study period i.e. 1991 onward. Therefore, no appreciable change was observed in snow metamorphic regime and thus no change in albedo was observed.

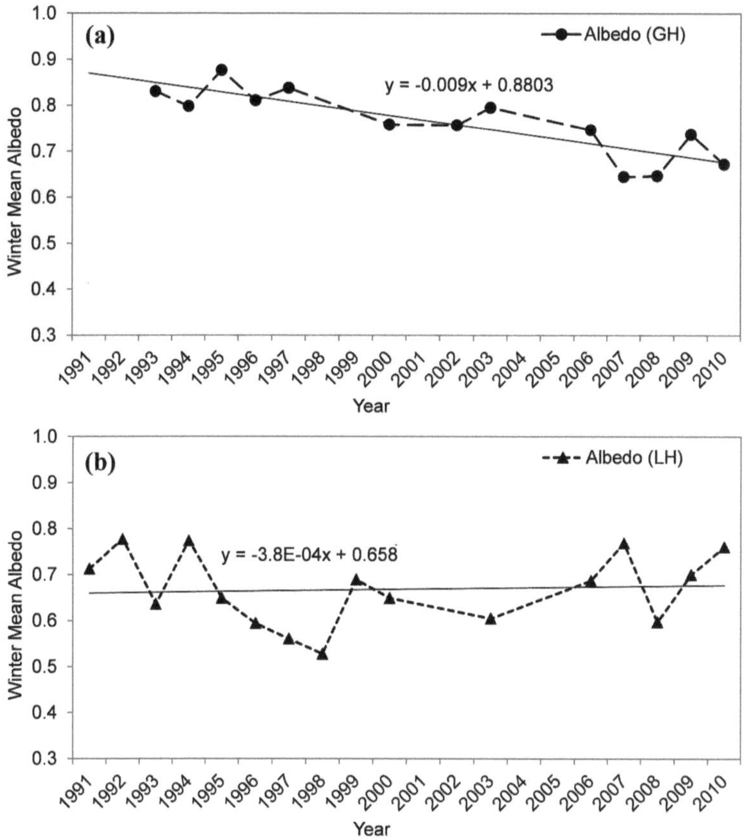

Fig. 7 Temporal variations in winter mean albedo **a** GH and **b** LH (Negi et al. 2017)

3.2 Climate Observations in Recent Years (2011–2014)

3.2.1 Snow Cover Area

The winter season snow cover area (SCA) maps generated for overall NWH suggest the insignificant increasing trend of SCA between 2001 and 2014 (Fig. 8a). Further the SCA trends of different NHW zones were found to be insignificantly increasing (Fig. 8b). The snow cover mapping study has been carried out by Gurung et al. (2011) for Hindu Kush Himalaya (HKH) for a period of 2000–2010. They suggested that overall increase in inter-annual SCA for western HKH and insignificant declining trend for winter SCA. Immerzeel et al. (2009) also reported no significant snow cover trends over Himalaya between 2000 and 2008. Singh et al. (2014) have recently reported a unique increasing trend in snow cover between 2000 and 2011 over Indus basin (our study area NWH is the part of Indus basin). Thus, the present study of recent 14 years is in agreement of increasing winter SCA over NWH.

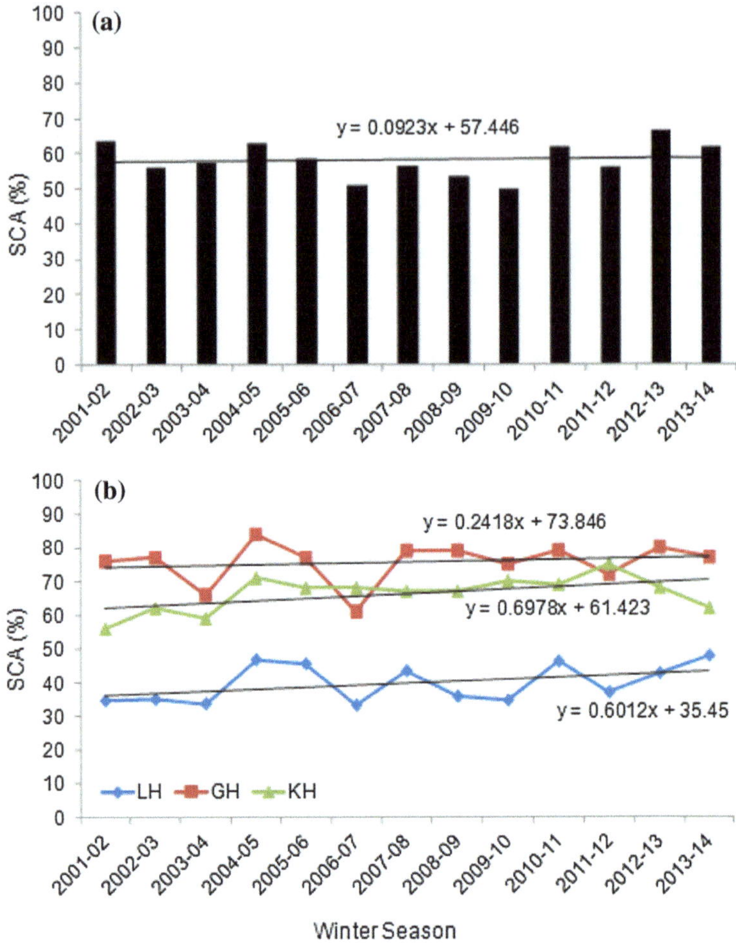

Fig. 8 Temporal variations in winter snow cover area **a** overall in NWH and **b** in different zones of NWH

3.2.2 Snow Albedo and Winter Mean Temperature

The overall winter snow albedo from MODIS over different Himalayan ranges for a period 2001–2014 is shown in Fig. 9a. The retrieved snow albedo values can clearly distinguish the different NWH ranges. The seasonal mean albedo using MODIS was found to be 0.80, 0.64 and 0.51 for KH, GH and LH ranges respectively. Thus the MODIS derived product is able to detect the spatial variations in albedo of different NWH ranges. An insignificant rise in the winter snow albedo in different NWH zones was observed from year 2001 onwards. These observations were found consistent with winter SCA analysis for recent decade.

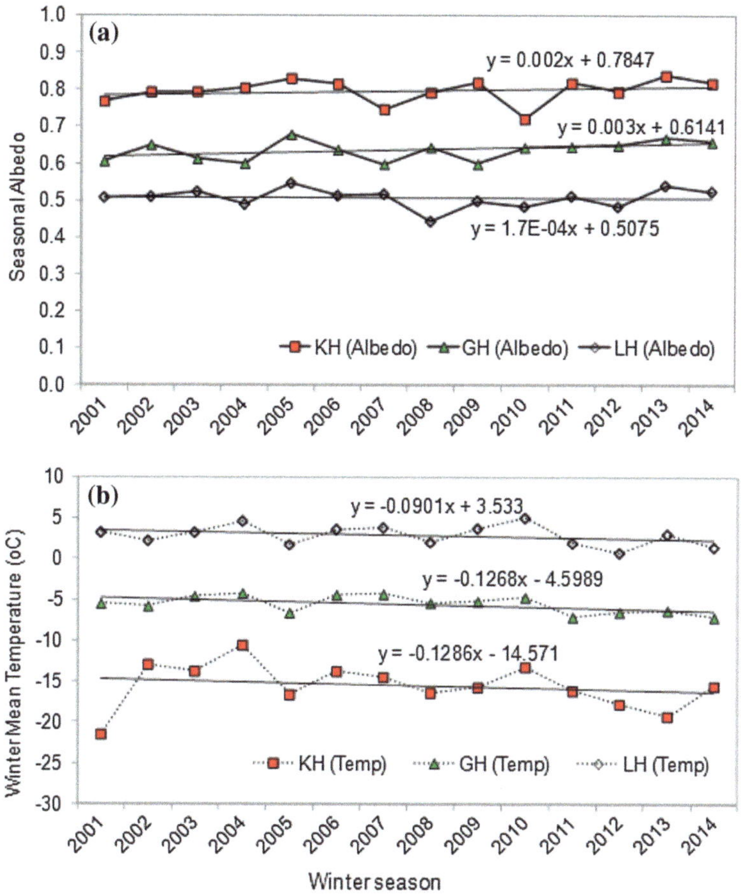

Fig. 9 **a** Temporal variations in winter mean albedo **b** temporal variations in winter mean temperatures over different zones of NWH (Negi et al. 2017)

Air temperature is one of the major factors for albedo variations as found from field data analysis, discussed in Sect. 3.1.3. Therefore, winter mean temperature pattern for the same duration was analysed and shown in Fig. 9b. The temperature patterns which were showing increasing trends for the period 1991–2011 (Fig. 4) are found to be in decreasing trend for the recent past 2001 onward. However, these decreasing trends were not found statistically significant (Fig. 9b and Table 2). The declining winter mean air temperature in recent years slowed down the rising trend of past two decades or can be termed as 'hiatus' in the winter mean temperature rise over NWH. As a result of this winter average albedo was found insignificant increasing over NW-Himalaya. The hiatus in global warming in recent decade has also been reported by many researchers (Guemas et al. 2013; Sillmann et al. 2014; Li et al. 2015).

4 Conclusions

Winter climate observations have important role for accumulation of snow and available snow mass at the end of accumulation for the remaining year. The past two decades (1991–2011) study that the increase in winter mean temperatures and decrease in albedo suggests snow-albedo feedback is a contributing factor in the amplified warming over GH region in Himalaya. Similarly increasing trends in SCA and albedo in recent years (2001 onward) support the declining trends in winter mean temperatures in all the zones of NWH. The results of SCA variations in NWH during recent 14 years are in agreement with the studies reported by others.

As albedo is a major forcing parameter for surface energy exchanges and hence, it is an important component for glacier mass balance. The albedo of snow covered glacier has implication on the overall thermal state of the snow pack even during the accumulation periods. As a result, the energy stored within the snow pack or glacier during winter may contribute in the accelerated ablation and vice-versa. The long-term rate of glacier retreat for approximately a period of 40 years (1960–2000) in Himalaya shows a significant loss of glacial length (Kulkarni and Karyakarte, 2014). But the recent short period study of glacier retreat (2001 onward) suggest that most of the glaciers are in a steady state compared to the rate of retreat prior to 2001 (Bahuguna et al. 2014). Here the results of our findings of winter climate, SCA and albedo study over NHW that the increase in temperature and decrease in albedo with respect to year 1991, and decrease in temperature and increase in SCA and albedo after 2001, support the recently reported findings of Himalayan glaciers variations in different duration. In addition significant increasing trend of precipitation mainly in LH and GH may also be a contributory factor for steady state or declining rate of retreat in glacier variation.

Acknowledgements The authors are thankful to scientists and technical staff of Snow and Avalanche Study Establishment (SASE), India, for collecting field data at remote locations of NWH. The authors are grateful to NSIDC to make available the MODIS snow products. This work was carried out under DRDO project 'Him Parivartan'.

References

Aoki T, Hachikubo AA, Hori M (2003) Effects of snow physical parameters on shortwave broadband albedos. J Geophys Res 108:4616

Bahuguna IM, Rathore BP, Brahmbhatt R, Sharma M, Dhar S, Randhawa SS, Kumar K, Romshoo S, Shah RD, Ganjoo RK, Ajai (2014) Are the Himalayan glaciers retreating? Curr Sci 106:1008–1013

Bhambri R, Bolch T, Kawishwar P, Dobhal DP, Srivastava D, Pratap B (2013) Heterogeneity in glacier response in the upper Shyok Valley, Northeast Karakoram. The Cryosphere 7: 1385–1398

Bhutiyani MR, Kale VS, Pawar NJ (2007) Long-term Trends in maximum, minimum and mean annual air temperatures across the North western Himalaya during the twentieth century. Clim Change 85(1–2):159–177

Bolch T et al (2012) The state and fate of Himalayan glaciers. Sci 336:310–314

Braganza K, Karoly DJ, Arblaster JM (2004) Diurnal temperature range as an index of global climate hange during the twentieth century. Geophys Res Lett 31. doi:10.1029/2004GL019998

Colbeck SC (1982) An overview of seasonal snow metamorphism. Rev Geophys Space Phys 20:45–61

Dash SK, Hunt JCR (2007) Variability of climate change in India. Curr Sci 93(6):782–788

Dash SK, Jenamani RK, Kalsi SR, Panda SK (2007) Some evidence of climate change in twentieth-century India. Clim Change 85(3–4):299–321

Dimri AP, Ganju A (2007) Wintertime seasonal scale simulation over Western Himalaya using RegCM3. Pure Appl Geophys 164(8–9):1733–1746

Easterling DR et al (1997) Maximum and minimum temperature trends for the globe. Sci 277: 364–367

Fowler HJ, Archer DR (2006) Conflicting signals of climatic change in the upper indus basin. J Clim 19(17):4276–4293

Guemas V, Reyes FJD, Burillo IA, Asif M (2013) Retrospective prediction of the global warming slowdown in the Past decade. Nat Climate Change 1–5

Gurung DR, Kulkarni AV, Giriraj A, Aung KS, Shresthaand B, Srinivasan J (2011) Changes in seasonal snow cover in Hindu Kush-Himalayan Region. Cryosphere Discuss 5:755–777

Hewitt K (2005) The Karakoram Anomaly? glacier expansion and the 'Elevation Effect', Karakoram Himalaya. Mt Res Dev 25(4):332–340

Immerzeel WW, Droogers P, de Jong SM, Bierkens MFP (2009) Large-scale monitoring of snow cover and runoff simulation in Himalayan River Basins using remote sensing. Remote Sens Environ 113:40–49

Jhajharia D, Singh VP (2011) Trends in temperature, diurnal temperature range and sunshine duration in Northeast India. Intern J Climatol 31:1353–1367

Klein AG, Stroeve J (2002) Development and validation of a snow albedo algorithm for the MODIS instrument. Ann Glaciol 34:45–52

Kothawale DR, Rupa Kumar K, (2005) On the recent changes in surface temperature trends over India. Geophys Res Lett 32(18):L18714

Kulkarni AV, Karyakarte Y (2014) Observed changes in Himalayan glaciers. Curr Sci 106 (2):237–244

Kumar Rupa, Krishna Kumar K, Pant GB (1994) Diurnal asymmetry of surface temperature trends over India. Geophys Res Lett 21(8):677–680

Li C, Stevens B, Marotzke J (2015) Eurasian winter cooling in the warming hiatus of 1998–2012. Geophys Res Lett 42:8131–8139

Ming J, Du Z, Xiao C, Xu X, Zhang D (2012) Darkening of the Mid-Himalaya glaciers since 2000 and the potential causes. Environ Res Lett 7:014–021

Negi HS, Kulkarni AV, Semwal BS (2009) Estimation of snow cover distribution in Beas basin, Indian Himalaya using satellite data and ground measurements. J Earth Sys Sci 118:525–538

Negi HS, Datt P, Thakur NK, Ganju A, Bhatia VK, Vinay KG (2017) Observed spatio-temporal changes of winter snow albedo over the north-west Himalaya. Int J Climatol 37(5):2304–2317

Pisharoty P, Desai BN (1956) Western disturbances and Indian weather. Indian J Meteorol Geophys 7:333–338

Qian Y, Flanner MG, Leung LR, Wang W (2011) Sensitivity studies on the impacts of Tibetan Plateau snowpack pollution on the Asian hydrological cycle and monsoon climate. Atmos Chem Phys 11:1929–1948

Roy SS, Balling RC (2005) Analysis of trends in maximum and minimum temperature, diurnal temperature range, and cloud cover over India. Geophys Res Lett 32(12):L12702

Sharma SS., Ganju A (2000) Complexities of avalanche forecasting in Western Himalaya—An Overview. Cold Reg Sci Technol 31:95–102. 31:L17201. doi:10.1029/2004GL02020

Shekhar MS, Chand H, Kumar S, Srinivasan K, Ganju A (2010) Climate change studies in the Western Himalaya. Ann Glaciol 51(54):105–112

Sillmann J, Donat MG, Fyfe JC, Zwiers FW (2014) Observed and simulated temperature extremes during the recent warming Hiatus. Environ Res Lett 9:1–8

Singh P, Kumar V, Thomas T, Arora M (2008) Basin-wide assessment of temperature trends in northwest and central India. Hydrol Sci J 53(2):421–433

Singh SK, Rathore BP, Bahuguna IM, Ajay (2014) Snow cover variability in the Himalayan-Tibetan Reg. Int J Climatol 34:446–452

Vose RS, Easterling DR, Gleason B (2005) Maximum and minimum temperature trends for the globe: an update through 2004. Geophys Res Lett 32:L23822

Yadav RR, Park WK, Singh J, Dubey B 2004. Do the Western Himalayas defy global warming? Geophys Res Lett 31(17). doi:10.1029/2004GL02020

Regional Climate Changes Over Hindukush-Karakoram-Himalaya Region

D. Kumar, A. Choudhary and A.P. Dimri

Abstract The Hindukush-Karakoram Himalaya, one of the largest mountain chains of the world with some of the highest peaks, comprises of regions of diverse climatic and hydrological regimes. In line with the global temperature rise, the model data sets, in spite of its being scare and not highly reliable, show a significant warming trend over the western parts of the Himalaya, while in other parts this positive trend is not significant, statistically. The former region has also recorded a positive trend of precipitation in the recent past in both the June to September summer monsoons and as well as in the winter (December to February) season; corresponding to the increase in the snow cover over some parts. A general decrease in the glacial mass has been reported by many studies which show that glaciers have retreated and lost their mass. However, in the central Karakoram, snow cover has increased in recent times (also called Karakoram anomaly).

Keywords Precipitation · Mass balance · Warming

1 Introduction

The Hindukush-Karakoram-Himalaya (HKKH) region extends up to 3500 km from north to south with Afghanistan in the west and Myanmar in the east. It includes Nepal and Bhutan as a whole along with the mountainous regions from Afghanistan, Bangladesh, China, Myanmar, Pakistan and India; thus traversing across eight countries. Again the HKKH region can be further geographically divided into two different sub-regions in which, first being the one having rugged terrain with altitude varying rapidly and the other having a high plateau type landform with less ruggedness (Singh et al. 2011). The first region with varying altitude contains almost all types of forest available on the earth while the other one has vast grasslands with comparatively flatter terrain covering area twice as much as

D. Kumar · A. Choudhary · A.P. Dimri (✉)
School of Environmental Sciences, Jawaharlal Nehru University, New Delhi, India
e-mail: apdimri@hotmail.com

P.S. Goel et al. (eds.), *Science and Geopolitics of The White World*,
DOI 10.1007/978-3-319-57765-4_11

the first region. The HKKH region comprises the biggest mountain chains of earth with largest ice mass extraneous to the Polar Regions. Due to its geographical setting, it gives rise to one of the most dynamic, complex, and fragile mountain ecosystems comprehending diverse climatic, hydrological and ecological traits in its premises. The region is appreciated not only for its multitudinous landscape beauty; but also for their role in shaping up the global atmospheric circulation, the peculiar hydrological cycle, enormous amount of ecosystem services, and available water resources. Figure 1 shows the spatial extent and geographical spread of Hindukush-Karakoram-Himalaya region.

The complex topography in the HKKH region has its own implications in shaping the climate over the area; thereby giving rise to different sub-regimes of climate in the same region. A variety of climatic conditions ranging from tropical to alpine can be found in the region; distributed in respect to the changing altitude (Kulkarni et al. 2013). Figure 2 shows the distribution of topography in the Himalayan and Tibetan region. The summer precipitation is relatively weaker in this region as the summer monsoon weakens from the east to west direction thus being unable to penetrate through the Karakoram ranges, thus giving rise to the decline in the same direction (Miller et al. 2012). The gradient of annual summer precipitation is such strong that the eastern region receives 10 times more precipitation (3000 mm) than the western part (300 mm) (Immerzeel et al. 2009). Singh et al. (2011) reported that eastern part receives almost 80% of its annual precipitation from the monsoonal season while for the western region encompassing northern Afghanistan and Pakistan, this share is merely 30%. In the western part, most of the precipitation (approx. 70%) is contributed by the westerly influences

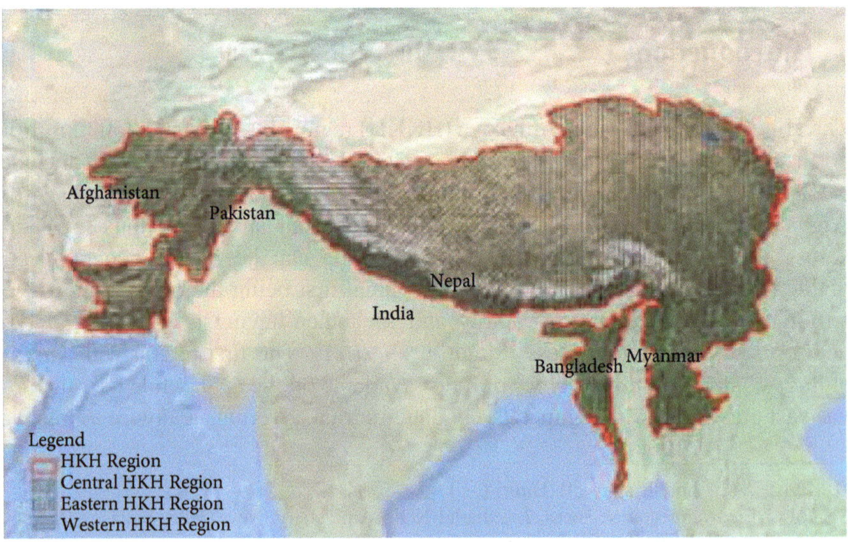

Fig. 1 Extent of the Hindu Kush Himalaya region. *Source* Gurung et al. (2011)

(Rajbhandari et al. 2015). Orography plays a very important role in determining the spatial distribution of the precipitation over a region such as Himalaya. Ghimire et al. (2015) have also shown the distribution of precipitation with respect to the elevation in the Himalayan region, which clearly shows that in the monsoon season, lower reaches of Himalaya receives higher amount of precipitation as compared to the higher elevation as indicated from the observation for the period 1970–2005 (Fig. 3). Such kind of variation in the precipitation patterns gives rise to a wide variety of the hydrological regimes in addition to different climatic sub-regimes. Attributed to this property, therefore, different parts of the region may respond differently to the climatic changes. Also, due to the complex topography, extreme variations of the altitude and inaccessibility and other technical and maintenance problems; the hydro meteorological data over the HKKH regions are scarce and the overall region is covered by only few weather stations (Singh et al. 2011). The region is endued with diverse flora and fauna with a rich gene pool attributed to the different climatic regimes which makes this ecologically sensitive ecosystem, important on a global scale. Four of the major global biodiversity hotspots are hosted by the region which includes- the Himalayas, Indo-Burma, Mountains of South-West China, and Mountains of Central Asia (Mittermeier et al. 2005).

The HKHH region acts as a storehouse for a large number of glaciers and seasonal snow thus also known as the third pole of the globe. Many rivers originate from this region which serve as a lifeline for the population downstream in many countries; as the water supplied by the glacier melting and snow, support maintain the flow in these rivers (Gurung et al. 2011). In the varying climatic regimes over the region, it has been estimated that the HKKH system hosts approx. 60,000 sq km of ice bodies including glaciers, glacierets and perennial surface ice (Fujita and Nuimura 2011; Kääb et al. 2012) providing water to the stream systems flowing

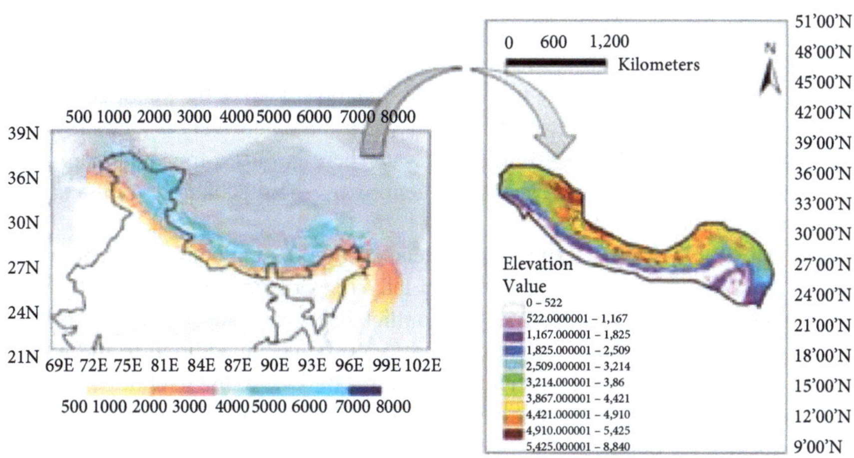

Fig. 2 Topography (meter) over **a** Himalaya and Tibetan region (*grey shaded*) **b** Study area. *Source* Ghimire et al. (2015)

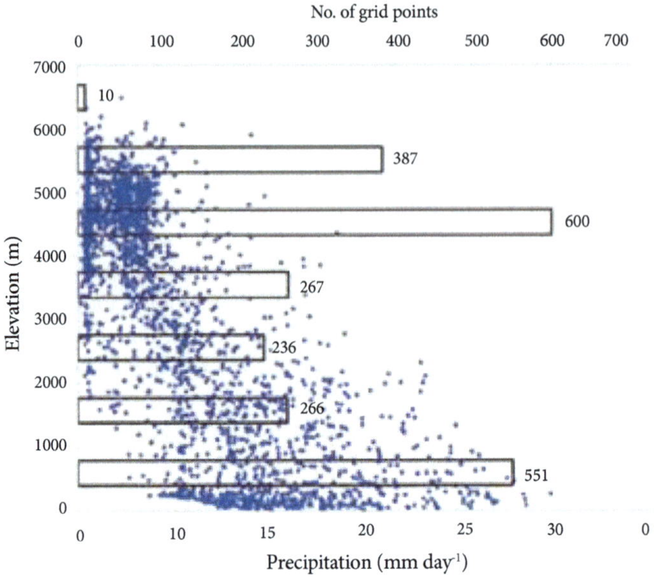

Fig. 3 Variation in observed precipitation (mm/day) with elevation. *Source* Ghimire et al. (2015)

downstream and support different activities such as agriculture, water supply for domestic and Industrial uses, Power production and many others.

The role of HKKH region in sustaining the population in the downstream regions can be understood from the fact that the water from the region replete the water resources for about 800 million people in Asia, especially during the crucial dry spells each year; thus known as the water tower of Asia (Immerzeel et al. 2010; Bolch et al. 2012). Major rivers draining the HKKH region are- Huanghe, Yangtze, Mekong, Salween, Brahmaputra, Ganges, Indus, Amu Darya, Tarim etc. The water supplied by these rivers drives the economy of the downstream countries and thus act as a survival base for a large number of population. Two of the major economies of the Asia as well as of the world—China and India extensively use this resource for various purposes. The distribution of water here is thus a transboundary issue which has multi-dimensional aspects influenced by socio-economic, political, and diplomatic implications.

The magnitude of these issues is significantly important at both micro and macro levels, as these are directly related to the daily lives of the dependent population. Such transboundary issues lead to the conflict between countries due to reasons specifically related to dams, irrigation, and hydro-power in addition to the regulation of the water flow, flood management and transit. Due to the livelihood interests and dependence over the water it is very common for the people to come across to regular local level conflicts.

These snow covered regions are thus important for the water resources in the downstream regions. It therefore becomes imperative to assess the impact of changing climate over these massive water towers.

This system has been replenishing the needs of the civilizations since quite a long time; and having noted it as one of the sensitive areas for the impacts of climate change, it is the need of the hour to ascertain and quantify the magnitude of the changes on a more local as well as regional scale. Climate change may have a serious impact over the cryosphere system. As these setup have immense influence over the regional climate system, it may be used as a sensitive indicator to climate change.

2 Regional Climate Change Over the Hindukush-Karakoram-Himalaya (HKKH) Region

2.1 Surface Temperature Changes

IPCC (2007) has concluded that there has been an increase in the global mean surface air temperature of the earth by 0.74 °C on an average in the last century, but this increase is not uniform at all the places. The fourth assessment report of the IPCC also predicts that it is likely for the earth to experience the increase in the mean temperature by the magnitude of 1.4–5.8 °C by the end of 21st century (IPCC 2007).

Bochiolla and Diolaiuti (2013) have analyzed the climate over some of the stations in the upper HKKH region and found that the maximum temperature regimes over the stations are positively correlated against the global anomaly of temperature significantly; while the minimum temperature is correlated negatively in a non-significant manner. This confirms the signature of the globally changing climate over the temperature climatology on the region as indicated from the above study for the period of 1980–2009. Same authors have reported that the maximum and minimum temperatures over the region have negative correlation with the NAO (North Atlantic Oscillation).

The Fig. 4 shows spatial distribution of trends of mean surface temperature over the period of 1970–2005 as simulated by regional climate model REMO with forcing from MPI-ESM-LR global model. The panels depict (a) the changes in the JJAS (June, July, August, and September) and (b) DJF (December, January, February) season over the Himalayan region. In the JJAS season, an evident cooling trend can be noticed over the western parts of the Karakoram-Himalaya; for the mentioned period, which somehow corroborates the Karakoram anomaly as reported by Hewitt (2005) thereby defying the global warming. However along the foothills of some of the central as well as the eastern Himalaya, a little greater warming trend is observed as compared to the remaining parts, which is sparsely distributed.

On the other hand, considering the DJF season a higher warming trend could be seen from the figure on the left for almost all the regions. In the western Himalaya,

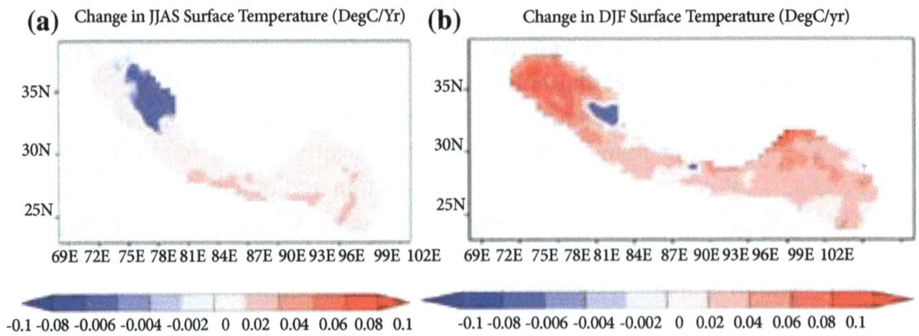

Fig. 4 Trend of **a** JJAS and **b** DJF mean surface temperature (°C/year), 1970–2005. *Source* Author

winters have been found to warm with a greater rate in all the parts except a smaller patch of the Karakoram-Himalaya where possibly the Karakoram anomaly neutralizes the warming trend a bit. Here it is important to notice that in the uppermost part of the eastern Himalaya winter time temperature is increasing at a faster rate as compared to the foothills. Also, in the central parts of the Himalaya, warming is pronounced in the winter season. These changes may have certain implications in ruffling the peculiar local climatic features prevailing in the area. A general increase of winter temperature in the upper parts of the Indus basin in the winter time temperature (approx. +0.6 °C) with a simultaneous cooling (roughly −1 °C) in the summer season for some of the stations have also been reported by Fowler and Archer (2005). While talking about the western Himalaya, Dimri and Dash (2012) stated that for the period of 1975–2006, the increase in the average temperature was found to be of the order of +0.6–1.3 °C. Shrestha and Devkota (2010) have also claimed that the annual mean temperature over the eastern Himalayan region have been increasing at a rate of 0.01 °C/year or more. Same authors also suggests that the warming trends are more pronounced during the winter months (i.e. December to February) at a rate of almost 0.015 °C more than the annual warming rate; and the regions with higher altitudes have undergone more warming during the same season for the period of 1975–2000.

Using these observations, similar results are reproduced by the REMO model simulations as shown in Fig. 5(b). Mishra (2015) has suggested all the observed data sets concord that a statistically significant warming trend over more than 40% area of the upper Indus basin has been observed for the period 1973–2007. In the same study, ensemble mean of the observed data sets indicates significant warming trends over the upper and lower Indus basin as well as the Ganga and Brahmaputra.

A study based on CMIP5 multi model projections by Chaturvedi et al. (2014) suggests that in the projected future climate over the Karakoram-Himalaya region; the mean temperature may increase by 2.36 °C (RCP 2.6) and 5.51 °C (RCP 8.5) when compared to the pre-industrial period (i.e. 1880s, averaged over 1861–1900).

Figure 6(a) describes the time series of the temperature anomalies from the CMIP5 multi model simulations for the past as well as the future. It is noticeable, that in the

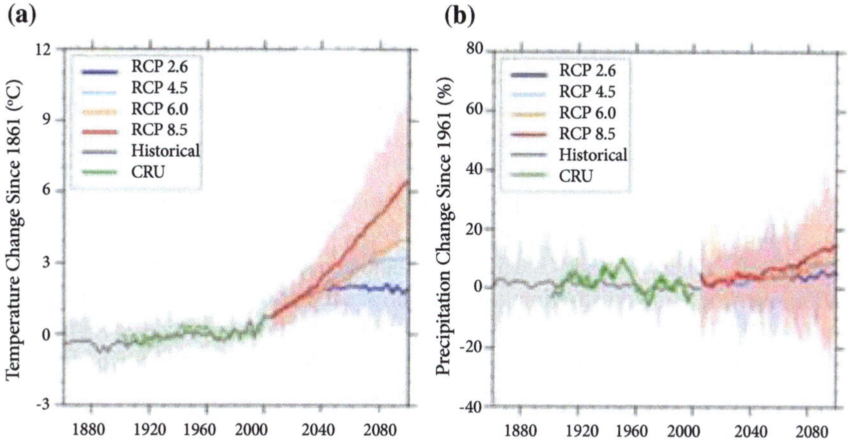

Fig. 5 Time series of **a** temperature anomaly b precipitation anomaly, 1861–2099. *Note* W.R.T base period of 1961–from CMIP5 multi-model simulations for different RCPs averaged over the Karakoram-Himalaya region. *Shaded region* represents the ranges of projections from 21 models from CMIP5 for each year. For each RCP, ensemble averages are shown in *thick colored lines*. *Source* Chaturvedi et al. (2014)

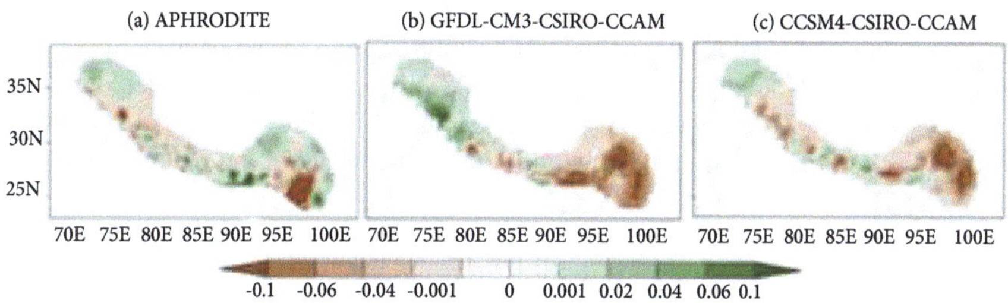

Fig. 6 JJAS precipitation trend (mm/year) for **a** Observation **b** GFDL-CM3-CSIRO-CCAM model experiment and **c** CCSM4-CSIRO-CCAM, 1970–2005. *Source* Ghimire et al. (2015)

past the temperature has seen an increasing trend over the Karakoram-Himalaya region from 1961–1990 average as shown by the ensemble of models as well as the observations. The projections by the ensemble of models show that the trend will continue to rise by the end of 21st century for all the scenarios with the sharpest rise seen in the most greenhouse gas (GHG) intensive RCP 8.5 scenario.

2.2 Precipitation

As stated in earlier sections as well, the HKKH region receives precipitation from two major sources: Monsoonal precipitation and western disturbances. Due to altitudinal effects, the precipitation is received both in the form of rain as well as snow.

Talking of the long term trend of the precipitation and using observed data; Palazzi et al. (2013) have shown that for the Himalayan region statistically decreasing trend of precipitation exists over the period of 1951–2007; while for the winter months none of the data sets used in the study give significant trend for Hindukush-Karakoram region. A study based on the station data over the northwest Himalayan region by Bhutiyani et al. (2010) suggests the statistically significant decreasing trend in monsoon and overall precipitation; but insignificant trend in the wintertime precipitation for the period of 1866–2006.

Similarly, Ghimire et al. (2015) have calculated the trend for the JJAS precipitation over the Hindukush-Himalaya region for the period of 1970–2005 using the APHRODITE data as observation and CORDEX model experiments. Figure 7(a) shows the spatial trend of the precipitation for the observation data in which mixed trend over the central Himalayan region could be observed. The upper reaches of the western Himalayan regions have been found to receive increasing precipitation over the period, as compared to the foothill regions. On the other hand, the decreasing trend in the lower elevations of the eastern Himalaya can also be seen in the same figure.

So besides the general perception, that the western Himalayan region receives less precipitation in monsoonal season, as stated in earlier sections, we may infer that over the period of time it may be possible that the spatial distribution of the monsoonal precipitation has changed in the Hindukush-Himalaya region.

Again, Fig. 7(b) and (c) depict the trend captured from the CORDEX South Asia model experiments, in which the CSIRO-CCAM regional model has been used with GFDL-CM3 and CCSM4 global model as initial forcing to them respectively. These two experiments are able to capture the spatial trend of the precipitation over

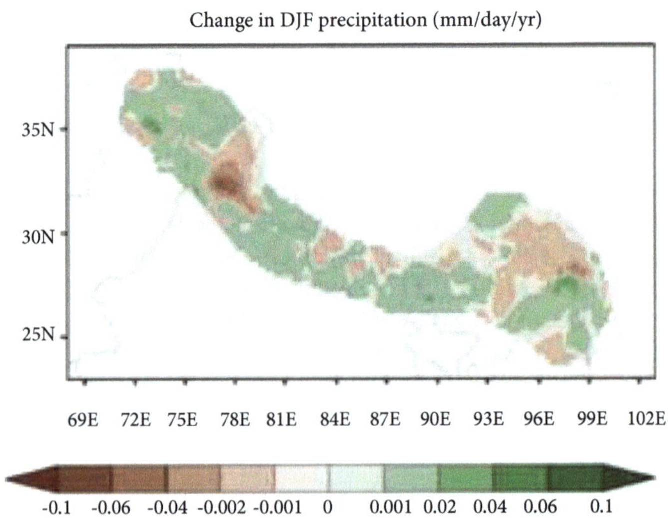

Fig. 7 DJF season for the period, 1970–2005. *Source* Author

the Himalayan region except for the upper reaches of the eastern Himalaya. The CCAM model with GFDL-CM3 forcing is able to capture the spatial variability of trend over the western as well as central Himalaya in a good manner. Both the experiments simulate similar kind of trend for the eastern Himalaya with different forcing, which may be due to the internal model physics or the initial forcing. Bochiolla and Diolaiuti (2013) have shown that in the upper Karakoram, mostly statistically non-significant changes in the precipitation climatology have been observed for the period 1980–2009. For a few stations of the study such as Chitral-Hindukush region and Northwest Karakoram, or Gilgit area, increase in the precipitation has been noticed; while for the western Himalaya, in Kotli region, decrease has been reported at the same time. However, except for some regions like Chitral and Gilgit areas, where slight increase in maximum precipitation has been reported; the maximum precipitation has remained unchanged in most of the cases.

Recently, observations based on APHRODITE shows that the monsoonal mean precipitation over the Himalayan region has decreased in the recent past with the rate of -0.006 mm day^{-1} year^{-1} on an average, during the period (1970–2005); though this trend is not statistically significant. This also supports the findings of Mukherjee et al. (2014); which claims that the trends calculated using APHRODITE data are not significant for most of the parts of the Himalaya. Some other studies e.g. Singh and Sen Roy (2002) for the Beas basin and Kumar and Jain (2010) for the Qazigund and Kukarnag of Kashmir, have also reported decreasing trends over the parts of western Himalayan region. In Fig. 8, we have shown the observed trends of precipitation for the winter season using APHRODITE data set (Yatagai et al. 2009). The figure depicts that in most of the parts of the western Himalaya and Karakoram region precipitation is positive, while the wintertime precipitation in maximum parts of eastern Himalaya are decreasing. This may have implication in the dynamics of mass balance of glaciers and water resources of the region which would be discussed in detail in subsequent sections. Again, in overall central Himalayan region, in the same figure the precipitation trends are found to be increasing. The rate of increase in seasonal wintertime precipitation at places is as high as 0.02 mm/day/year.

Chaturvedi et al. (2014) have shown the changes in precipitation since 1861 as well as the possible changes in precipitation (Fig. 6b) in the future RCP scenarios using CMIP5 multi model simulations for the entire Karakoram-Himalaya region. This study shows that using the CRU observations, no clear trend in precipitation changes was inferred for the past as well as the present climate; but in the most extensive warming scenario of RCP 8.5 the annual mean precipitation may change by the end of the 21st century. Further, the increase in the precipitation may range from 0.6 to 1.6% by the 2030 s under RCP 2.6 and RCP 8.5 scenario respectively and by the period 2080s this increase may reach up to 2.6–8.5% for the same scenarios as related to the pre-industrial period.

Besides discussing the trends of precipitation over the HKKH region, the extent of spatial variability in seasonal mean precipitation is also important to explore the changes in the spatial variability of the precipitation. Here in Fig. 9, we show the standard deviation of the precipitation for the monsoonal month using observed as

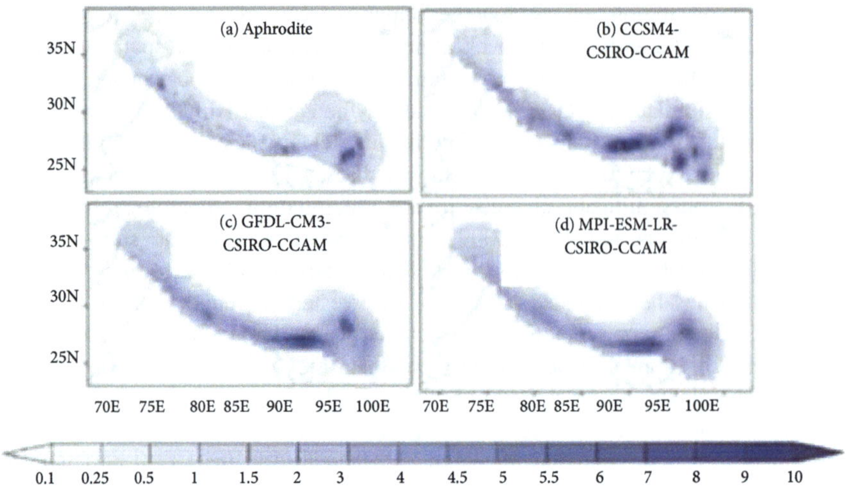

Fig. 8 Standard deviation (mm/day) for JJAS precipitation for **a** Observation **b** CCSM4-CSIRO-CCAM experiment **c** GFDL-CM3-CSIRO-CCAM experiment and **d** MPI-ESM-LR-CISRO-CCAM experiment under CORDEX South Asia framework, 1970–2005. *Source* Ghimire et al. (2015)

well as the model data sets. This figure presents the variability in the seasonal mean monsoonal precipitation in the recent past (1970–2005). As we can see the observation (a) and model (b–d) all show a higher variability along the foothills in the eastern Himalaya compared to the westernmost stretch of the study region. These point towards the sensitive state of the climate in the former. Eventually the region of higher variability i.e. the Eastern Himalaya is also the region which receives higher precipitation in the monsoon season. This suggests that it may experience a higher inter-annual fluctuation in the seasonal mean precipitation of up to 5 mm/day.

The spatial distribution of the precipitation variability of the DJF season has been presented in the Fig. 10. The magnitude of variability in the observed precipitation is comparatively less throughout the region when compared to the JJAS season. However in JJAS season along the foothills of the western Himalaya; variability is higher due to the reason that this area receives higher amount of precipitation during winter season due to the influence of western disturbances.

As discussed earlier, the trends of precipitation over the eastern Himalaya has been found to be negative for the period 1970–2005. In this respect we have calculated the changes in the cloud area fraction for the HKKH region as simulated from the REMO-2009 model under the CORDEX South Asia framework. Figure 11(a) represents the changes for the monsoonal season, in which we can see that the eastern Himalaya region where the precipitation trends were negative; apparently over the same region the changes in cloud area fraction are also negative. This also holds true for the western Himalaya where the precipitation trends were increasing, as the cloud area fraction is increasing over the same area. In the central

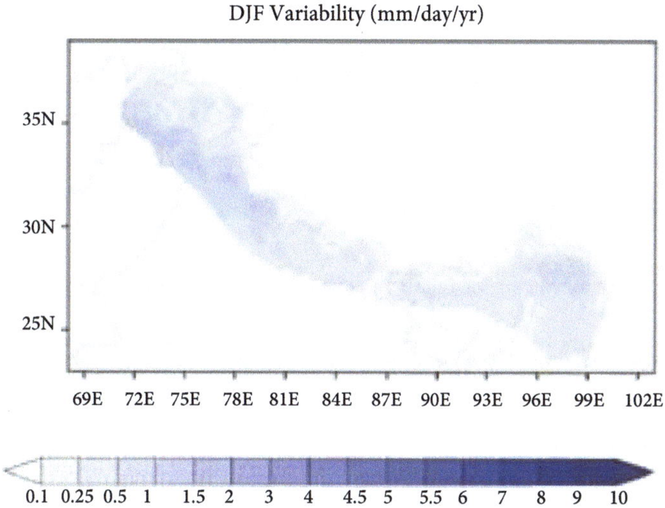

Fig. 9 DJF season over the HKKH region 1970–2005. *Source* Author

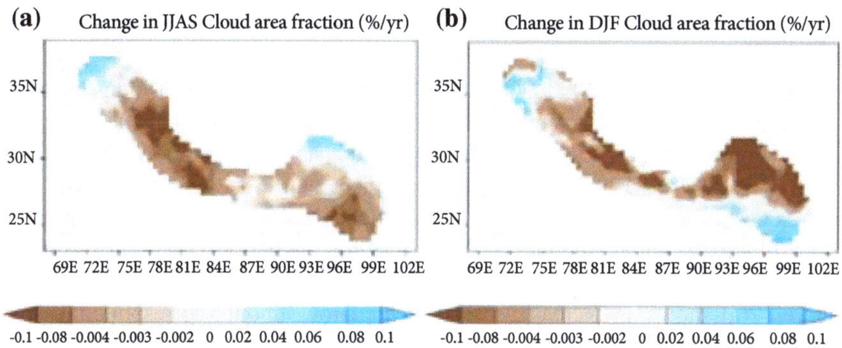

Fig. 10 Changes in the cloud area fraction for **a** JJAS and **b** DJF season over the HKKH region, 1970–2005 as simulated by the Remo-2009 model. *Source* Author

Himalayan region for monsoonal season, negative trends of cloud area fraction could also be seen.

2.3 Glacier and Snow Cover Changes

As per the estimate from the satellite data from 2000 to 2010; the snow covered area over the Hindukush-Karakoram-Himalaya is 0.76 million km^2, representing almost 18.23% of the geographical area of the region (Gurung et al. 2011). The region is

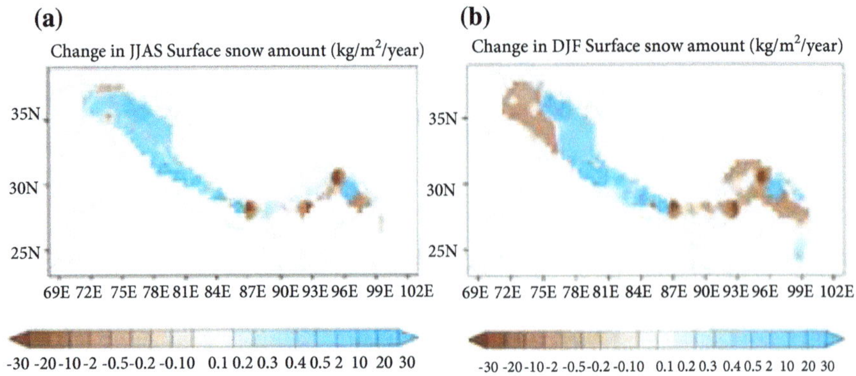

Fig. 11 Spatial distribution of trends of surface snow amount (kg/m^2/year) for **a** JJAS and **b** DJF season, 1970–2005. *Source* Author

characterized by a large climatic variability attributed to the differences in the altitude, which in turn results into the large spatial variation in the snow cover throughout the region (Immerzeel et al. 2009; Pu et al. 2007). In the HKHH region, due to unavailability of data pertaining to snow cover for a long period, it is difficult to perform trend analysis in order to assess the changes over the area (Gurung et al. 2011).

A study based on the 2000–2008 snow cover data over the upper Indus basin and some other regions gives significant trend for the snow cover for winter over the region (Immerzeel et al. 2009). While some of the stations over the Tibetan plateau for the period 1957–1992 suggest a striking increasing trend in snow depth (Dahe et al. 2006). In order to show the changes in the snow cover, we have calculated the trends of surface snow amount (kg/sq m/year) for the period 1970–2005 using the output of REMO-2009 model under the CORDEX South Asia experiment. Figure 12(a) shows an increasing trend of the snow amount corresponding to the positive trends in the precipitation over the western Himalaya. Again conforming to the decreasing precipitation trends the eastern Himalayan region has experienced the decreasing trends of surface snow amount for the period. Further looking for the possible role of temperature changes over the region in the changes in snow amount, we can see from Fig. 12(b) that over the westernmost parts of the western Himalaya, where positive temperature changes have occurred in the recent past; a decrease in the overall surface snow amount is evident. This fact apparently holds true for the eastern Himalaya as well. However there is a homogenous increase in the snow cover over the central Himalayan region; which possibly has not responded much to the changes in the temperature as well as precipitation over the same region.

Gurung et al. (2011) have carried out the trend analysis for the snow cover over the entire HKH Himalaya and found that the decadal snow cover has significantly increased by +8.55 ± 1.70% over the region since 2000. For other regions of the Himalaya like eastern and central parts, such trends are not significant, statistically.

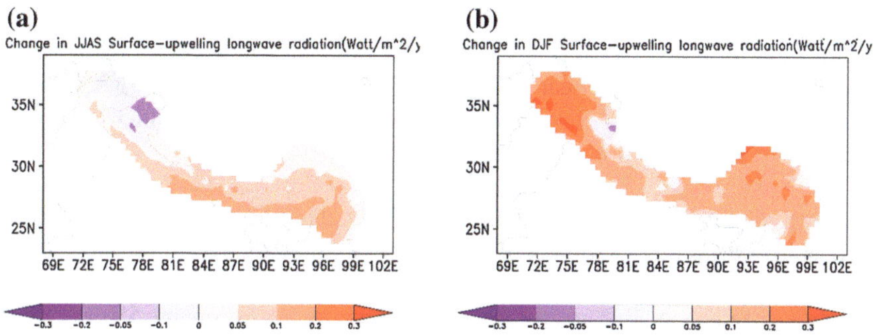

Fig. 12 Spatial distribution of trends of surface upwelling long wave radiation (watt/m^2/year) for **a** JJAS and **b** DJF Season, 1970–2005. *Source* Author

Table 1 Snow cover trends for Hindukush-Himalaya region, 2000–2010

	HKH region	Western HKH (%)	Central HKH (%)	Eastern HKH (%)
Spring	−1.04% ± 0.97	+4.04 ± 2.02	0.56 ± 2.67	−0.89 ± 1.23
Summer	−0.01% ± 0.46	+2.58 ± 0.74	+0.47 ± 0.65	+3.35 ± 1.62
Autumn	+5.63 ± 8.39	+3.69 ± 2.64	+15.53 ± 5.69	+7.74 ± 5.84
Winter	−0.54% ± 0.67	+5.08 ± 2.48	−3.68 ± 3.01	+1.61 ± 3.12
Annual	−1.25 ± 1.13%	+8.55 ± 1.70	+1.66% ± 2.26	+0.82 ± 2.50

Source Gurung et al. 2011
Note The significant trends are shown in boxes

Further in the same study, it has been pointed out that over this period in the western and Hindukush-Himalaya region the snow cover have changed positively but this change is non-significant (refer to Table 1). However, Menon et al. (2010) have reported that there has been a decline in the annual snow cover by—16% per decade during 1990–2000 over the Hindukush-Himalaya region.

Corresponding to the increases in the surface snow amount as explained above (Fig. 12), the regions where the snow covers are advancing in the JJAS season show decreasing trends of the surface upwelling long wave radiation (Fig. 12a). This may be due to the fact that, since the surface is covered with the snow which is having higher albedo; so the incoming downward shortwave radiation is reflected back. Since it is not absorbed by the surface, it does not return back as long wave radiation. Again for the DJF season, the regions with decreasing trend in surface snow amount shows increasing trend for the upward long wave radiation which in turn results into the increasing trend in surface temperature due to absorption as evident from Fig. 5(b). The rate of this positive change is comparatively higher in the foothills as well as the uppermost parts of the western Himalaya in particular for the DJF season.

Many studies have reported the changes in the glacier mass especially focusing the retreat over the recent past as a response to the climatic changes. Thakur et al. (1991)

have reported the average rate of recession for the Gangotri glacier, which was found to be of the order of ~ 18 m per year. Similarly in a study conducted by Dobhal et al. (1999), a retreat of approx. 586 m was reported for the Dobriani Bamak glacier in Garhwal Himalaya for the period of 1962–1997, which has caused the shifting of the snout of the glacier.

For the same region Matny (2000) reported the retreat for a single year (1998) of the order of 20 m, which is significantly higher than the average retreat rate of 16.5 m over the past 35 years for the same region. Similarly for the period of 1963–1997, Kulkarni et al. (2004) found the retreat of Janapa glacier by 696 m, for Jorya Garang—a retreat of 425 m, for Naradu Garang—a recession by 550 m, Bilare Bange—retreat by 90 m, Karu Garang—retreat by 800 m and Baspa Bamak—a retreat by 380 m respectively. In the recent past, the glacier retreat and fluctuations have been prominent, which has caused a significant loss in the glacier mass and area over the Hindukush-Himalaya region (Salerno et al. 2008; Bolch et al. 2012; Kääb et al. 2012). In the Karakoram region, frequent observations of the glacier advancement or slower retreat rates have been reported which is in contrast to that over the eastern Himalayan region. This behaviour is termed as Karakoram anomaly. However across the Himalayan region, in general, and most of the major land parts of the Asia, rapid decline in the glacial mass (Ageta and Higuchi 1984; Ageta 1996) is reported attributed to the global atmospheric warming (IPCC 2001; 2007). Vohra (1981) has suggested that the Siachen and Pindari glaciers have undergone the retreat at the rate of 31.5 m and 23.5 m respectively in the recent past. A study focusing over the glacial thinning over Jammu-Kashmir region suggested a maximum thinning rate of 0.66 ± 0.09 m/year (Kääb et al. 2012). In the Karakoram-Himalaya region, during 1997–2001; 33 glaciers have been observed to thicken (by approx. 5–20 m in the lower part) and/or advance or at least stagnate (Hewitt 2005). A very general understanding regarding climate change suggests that the extent of Himalayan glaciers are shrinking but at the same time in Karakoram more than 30 glaciers have expanded not because of the lower temperatures but due to the fact that the higher temperatures have caused the existing ice to move down slope faster.

Philip and Sah (2004) compared the snout of the Shaune Garang glacier from the topographic map of 1962 with the one in the field survey and concluded a vertical shift of 100 m and lateral shift of 1500 m in the snout during the period under consideration. Such observation may lead to a raw conclusion that the changes in the glacier mas are attributed to the global warming; which has affected the snow-glacier melt and runoff patterns over the region. A glacier expedition group in Nepal led by Asahi et al. (2006) measured the retreat in Khumbu and Shorang glaciers and concluded that the glacier in the Shorang area have retreated at a rate of 8 m/yr, while in the Khumbu glacier, the rate of retreat has been recorded as 5–10 m/yr. The same group has also concluded that the glacier retreat rate has accelerated since 1990 (Asahi et al. 2006). Another study from Nepal Himalaya found that during 1970–2000 period, loss of glacier area has been of the order of 5.88 or 0.2% per year in the Tamour Basin (Bajracharya et al. 2006).

3 Summary and Conclusions

Hindukush-Karakoram-Himalaya region is the largest mountain system of the world traversing across many countries. The complex topographical settings accompanied by the extreme variations in the elevation and aspect give rise to many peculiar climatic regimes from east to west across the region which behave differently. Besides the diversity in the climatic regimes a wide variety of hydrological regimes have also been found in the region; which in turn makes this system dynamic and sensitive. There has been a lot of discussion about the climatic changes in the recent past; in this respect it becomes imperative to assess the changes in such non-resilient and sensitive system like HKKH.

In this paper, we have tried to summarize the observed changes in the climate system; which have taken place in the recent past over the HKKH region using observation as well as the model data sets. Corresponding to the global temperature rise; a significant warming trend over the western parts of the Himalaya have been reported in the recent past. The other parts of the HKKH region are also dominated by this positive trend of the temperature but their trend is not significant statistically. This may have implications for the hydrological cycle of this area, known as third pole of the world. Again, studies report that significant decrease in the precipitation trend has been observed over the eastern Himalayan region, while for other parts in the same region, a non-significant trend is dominating. It has also been noticed that the western Himalaya has experienced a positive trend of precipitation in the recent past in both JJAS as well as the DJF season; corresponding to the simultaneous increase in the snow cover over some parts. However a general decrease in the glacial mass has been reported by many studies which show that glaciers have lost their mass while facing the menace of retreat. At the same time, for the central Karakoram part, snow cover has increased in recent times (also called Karakoram anomaly). The lack of data over the HKKH region has been a problem in order to evaluate the long term changes in the climatic, hydrological and other systems. As of now, the state of current knowledge on the effect of the climate is miniscule and vague. Thus in order to plan for the impact and adaptation studies, the availability of reliable long term observational as well as model data is important.

Considering the importance of this mountain system in terms of its role in sustaining a large population by providing the natural resources; many studies have observed and projected the impact of climatic changes over the region for the atmospheric and hydrological components. This in turn affects the flora and fauna for the same region; as it has been seen that many of the species are facing the threat of extinction.

Thus it becomes urgent to develop an understanding of the changes on a more regional as well as local scales over the HKKH region in order to minimize the socio-economic losses in the future changing climate. Moreover, there is an urgent need of more extensive studies with holistic approach to get the clear picture of the current status of the changing climate of the "Water tower of Asia".

References

Ageta Y, Higuchi K (1984) Estimation of mass balance components of a summer-accumulation type glacier in the Nepal Himalaya. Geogr Annaler. Ser A. Phys Geogr 66 A:249–255

Ageta Y (1996) Characteristics of mass balance of summer-accumulation type glaciers in the himalayas and Tibetan Plateau. Zeitschrift fur Gletscherkunde und Glazialgeologie 32:61–65

Asahi K, Kadota T, Naito N, Ageta Y (2006) Variations of small glaciers since the 1970s to 2004 in Khumbu and Shorang regions, eastern Nepal. Data Rep 4:109–136

Bajracharya SR, Mool PK (2006) Impact of global climate change from 1970s to 2000s on the glaciers and glacial lakes in Tamor Basin, eastern Nepal. Unpublished report for ICIMOD, Kathmandu

Bhutiyani MR, Kale VS, Pawar NJ (2010) Climate change and the precipitation variations in the northwestern Himalaya: 1866–2006. Int J Climatol 30(4):535–548

Bocchiola D, Diolaiuti G (2013) Recent (1980–2009) evidence of climate change in the upper Karakoram, Pakistan. Theoret Appl Climatol 113(3–4):611–641

Bolch T, Kulkarni A, Kääb A, Huggel C, Paul F, Cogley JG, Stoffel M (2012) The state and fate of Himalayan Glaciers. Science 336(6079):310–314

Chaturvedi RK, Kulkarni A, Karyakarte Y, Joshi J, Bala G (2014) Glacial mass balance changes in the Karakoram and Himalaya based on CMIP5 multi-model climate projections. Clim Change 123(2):315–328

Dahe Q, Shiyin L, Peiji L (2006) Snow cover distribution, variability, and response to climate change in western China. J Clim 19(9):1820–1833

Dimri AP, Dash SK (2012) Wintertime climatic trends in the western Himalayas. Clim Change 111(3–4):775–800

Dobhal DP, Gergan JT, Thayyen RJ (1999) Recession of Dokriani Glacier, Garhwal Himalaya'—An overview. A Himal Perspect, Symposia on snow, ice and glaciers in geological survey India, pp 30–33

Fujita K, Nuimura T (2011) Spatially heterogeneous wastage of Himalayan glaciers. Proc Natl Acad Sci 108(34):14011–14014

Fowler HJ, Archer DR (2005) Hydro-climatological variability in the upper indus basin and implications for water resources. Reg Hydrol Impacts Clim Change—Impact Assess Decis Mak 295:131–138

Gurung DR, Kulkarni AV, Giriraj A, Aung KS, Shrestha B, Srinivasan J (2011) Changes in seasonal snow cover in Hindu Kush-Himalayan region. Cryosphere Discuss 5(2):755–777

Ghimire S, Choudhary A, Dimri AP (2015) Assessment of the performance of CORDEX-South Asia experiments for monsoonal precipitation over the Himalayan Region during present climate: part I. Clim Dyn 1–24. doi:10.1007/s00382-015-2747-2

Hewitt K (2005) The Karakoram anomaly? glacier expansion and the 'Elevation Effect'. Karakoram Himalaya. Mountain Res Develop 25(4):332–340

Immerzeel WW, Droogers P, De Jong SM, Bierkens MFP (2009) Large-scale monitoring of snow cover and runoff simulation in Himalayan River Basins using remote sensing. Remote Sens Environ 113(1):40–49

Immerzeel WW, Van Beek LP, Bierkens MF (2010) Climate change will affect the asian water towers. Sci 328(5984):1382–1385

Intergovernmental Panel on Climate Change IPCC (2001) Climate Change, 2001, The scientific basis. Contribution of working group I to the third assessment report of the Intergovernmental panel on climate change. Cambridge University Press, pp 892

Intergovernmental Panel on Climate Change IPCC (2007) Climate change 2007: Synthesis Report. IN: Pachauri, RK, Reisinger A (eds) Contribution of working groups I, II and III to the fourth assessment report of the intergovernmental panel on climate change. Geneva, Switzerland, p 104

Kääb A, Berthier E, Nuth C, Gardelle J, Arnaud Y (2012) Contrasting patterns of early twenty-first-century glacier mass change in the Himalayas. Nat 488(7–412):495–498

Kulkarni AV, Rathore BP, Alex S (2004) Monitoring of glacial mass balance in the Baspa basin using accumulation area ratio method. Curr Sci 86(1):185–190

Kulkarni A, Patwardhan S, Kumar KK, Ashok K, Krishnan R (2013) Projected climate change in the Hindu Kush-Himalayan region by using the high-resolution regional climate model PRECIS. Mt Res Dev 33(2):142–151

Kumar V, Jain SK (2010) Trends in seasonal and annual rainfall and rainy days in Kashmir Valley in the last century. Quatern Int 212(1):64–69

Matny L (2000) Melting of Earth's ice cover reaches new high. In: Worldwatch NewsBrief, 06 Mar 2000

Menon S, Koch D, Beig G, Sahu S, Fasullo J, Orlikowsk D (2010) Atmospheric chemistry and physics black carbon aerosols and the third polar ice cap. S Atmos Chem Phys 10:4559–4571. doi:10.5194/acp-10-4559-2010. www.atmos-chem-phys.net/10/4559/2010/

Miller JD, Immerzeel WW, Rees G (2012) Climate change impacts on glacier hydrology and river discharge in the Hindu Kush-Himalayas: a synthesis of the scientific basis. Mt Res Dev 32(4): 461–467

Mishra V (2015) Climatic uncertainty in Himalayan water towers. J Geophys Res: Atmos 120(7): 2689–2705

Mittermeier RA, Gil RP, Hoffman M, Pilgrim J, Brooks T, Mittermeier CG, Lamoreux J, da Fonseca GAB (2005) Hotspots revisited: Earth's biologically richest and most endangered ecoregions. CEMEX, Mexico City (Mexico)

Mukherjee S, Joshi R, Prasad RC, Vishvakarma SC, Kumar K (2014) Summer monsoon rainfall trends in the Indian Himalayan region Theoretical and Applied Climatology 1–14

Palazzi E, Hardenberg J, Provenzale A (2013) Precipitation in the Hindu Kush Karakoram Himalaya: observations and future scenarios. J Geophys Res: Atmos 118(1):85–100

Philip G, Sah MP (2004) Mapping repeated surges and retread of glaciers using IRS-1C/1D data: a case study of Shaune Garang glacier, northwestern Himalaya. Int J Appl Earth Obs Geoinf 6(2):127–141

Pu Z, Xu L, Salomonson VV (2007) MODIS/Terra observed seasonal variations of snow cover over the Tibetan Plateau Geophys Res Lett 34(6)

Rajbhandari R, Shrestha AB, Kulkarni A, Patwardhan SK, Bajracharya SR (2015) Projected changes in climate over the Indus River Basin using a high resolution regional climate model (PRECIS). Clim Dyn 44(1–2):339–357

Salerno F, Buraschi E, Bruccoleri G, Tartari G, Smiraglia C (2008) Glacier surface-area changes in Sagarmatha National Park, Nepal, in the second half of the 20th century, by comparison of historical maps. J Glaciol 54(187):738–752

Shrestha AB, Devkota LP (2010) *Climate change in the Eastern Himalayas: observed trends and model projections*. International Centre for Integrated Mountain Development (ICIMOD) pp 1–20

Singh SP, Bassignana-Khadka I, Karky BS, Sharma E (2011) Climate change in the Hindu Kush-Himalayas: the state of current knowledge. International Centre for Integrated Mountain Development (ICIMOD)

Singh RB, Sen Roy S (2002) Climate variability and hydrological extremes in a Himalayan catchment. In: ERB and Northern European FRIEND Project 5 Conf., Slovakia

Thakur VC, et al. (1991) Report on Gaumukh. The snout of the Gangotri Glacier

Vohra CP (1981) Note on recent glaciological expedition in Himachal Pradesh. Geol Surv India 6:26–29 Special Publication

Yatagai A, Arakawa O, Kamiguchi K, Kawamoto H, Nodzu MI, Hamada A (2009) A 44-year daily gridded precipitation dataset for Asia based on a dense network of rain gauges. Sola 5:137–140

Current Status of Himalayan Cryosphere and Adjacent Mountains

Abul Amir Khan, N.C. Pant and Rasik Ravindra

Abstract Hindu Kush-Karakoram-Himalaya (HKKH) region represents one of the major non-polar cryosphere domains on the Earth. This region feeds three major rivers namely: the Indus, the Ganga and the Brahmaputra and supports a huge population of more than 1 billion people. There is wide variability and uncertainty in data on most aspects of this cryospheric domain. The behaviour of glacial melting in HKH region is highly heterogeneous with the highest negative mass balance in the eastern Himalaya, relatively less negative mass balance in the western Himalaya with positive mass balance in the Karakoram. The hydrological budget of the higher Himalayan rivers depends on the precipitation (snowfall and rainfall) but the available estimates on snow cover and rainfall are highly variable and in few cases appear to be unacceptable. Reported precipitation variability for the Indus basin is more than 250%, for the Ganga basin it is 100% and for the Brahmaputra basin the variability is more than 240%. The estimate on glacial cover and its volume in the Himalayan-Karakoram regions shows variability of more than 130 and 250% respectively. The available estimates on the glacial melt fraction also show high variability, for example for the Indus basin the variability is ~170%, for the Ganga basin it is ~300% and for the Brahmaputra basin the variability is more than 100%. The number of glaciers in the Himalaya and the adjacent mountains differ in the different glacier inventories. Similarly, published data on basin wise glaciated area varies from 300% for Indus basin, 200% for the Ganga basin and it is more than 450% for the Brahmaputra basin. The present work reviews current status of the Himalayan cryosphere.

Keywords Mass balance · Variability · Indus · Ganga · Brahmaputra · Glacial melt

---

A.A. Khan · N.C. Pant (✉)
Department of Geology, University of Delhi, New Delhi, India
e-mail: pantnc@gmail.com

R. Ravindra
Formerly with Ministry of Earth Sciences (MoES), New Delhi, India

P.S. Goel et al. (eds.), *Science and Geopolitics of The White World*,
DOI 10.1007/978-3-319-57765-4_12

1 Introduction

The term Cryosphere originally derived from the Greek word 'kryos' meaning Icy cold was introduced by Dobrowolski (1923). The cryosphere may be defined as all the frozen water and soil on the earth's surface including perennial and seasonal snow cover, glaciers, ice sheets, ice shelves, ice bergs, sea ice, fresh water ice, permafrost and ground ice. There are an estimate of 70 large ice caps and ~160,000 other non-ice sheet glaciers covering nearly 15 million km^2 of the surface of the Earth. There are about ~33 million km^3 of ice on the earth's surface which occupy about 10% of the earth's land surface area and holds roughly 77% of fresh water. More than 96% of glaciers exists in the polar regions (Meier and Bahr 1996; Dyurgerov and Meier 2005; Paul et al. 2009; Williams and Ferrigno 2012). The Hindu Kush-Himalayan region (HKH) has the largest concentration of snow and glaciers outside the polar regions and has been referred as the Third Pole (Dyhrenfurth 1955; Bajracharya et al. 2011). The HKKH region covers more than 4 million km^2 of mountainous regions and extends across eight countries namely Afghanistan, Pakistan, China, India, Nepal, Bhutan, Bangladesh and Myanmar. There are more than 54,000 glaciers in this region which cover an area of ~60,000 km^2 and comprise more than 6000 km^3 of ice reserves (Bajracharya et al. 2011). Snow cover represents the second largest component of the cryosphere (Barry and Gan 2011). The average snow covered area of this region is ~0.76 million km^2 which represents ~18% of the total geographical area of the region (Gurung et al. 2011).

Various components of the cryosphere are sensitive indicators of the global climatic changes documented by changes in the volumes or areas of glaciers, reduction in depth and duration of snow cover, decrease in the thickness and areal extent of the sea ice, reduction of the duration and extent of the lake and river ice and diminishing permafrost area (Williams and Ferrigno 2012). The water which is stored in the glacier or in ice sheet has a potential to raise sea level by 75.6 m. Most of this is ascribed to the Antarctic ice sheet (68 m), followed by Greenland ice sheet (7 m) and the remaining 0.6 m to the non-ice sheet glaciers (Williams and Ferrigno 2012). Thus, cryosphere constitute an important subset of the hydrosphere and plays a significant role in the global climate system as it is the second largest store house of water after Ocean (Barry 1987, 2002).

An important positive climatic feedback is the ice-albedo feedback mechanism. Fresh snow and ice has the higher albedo i.e. they can reflect a huge amount of incoming solar radiation back to space. The expansion of snow and ice cover on the earth surface increases the albedo and deceases the temperature, which in turn enables the snow and ice cover to expand further. Thus, lower temperatures lead to more ice growth and consequently higher albedo, which in turn cools the surface. This is called positive snow-ice feedback. But in the present scenario this effect is operating in an opposite direction as shrinkage of snow and ice cover is lowering the albedo and increasing the absorption of incoming solar radiation. Increased absorption of solar radiation raises the temperature which in turn further reduces the snow and the ice cover. On a regional or global scale, depending on the cloud cover, the snow-ice–albedo effects supplement climate sensitivity by about 25–40% (Barry and Gan 2011). Each component of the cryosphere contribute to short-term climate changes, with permafrost, ice shelves and ice sheets or ice caps also contributing to long term changes including the ice age cycles.

Cryosphere is a sensitive indicator of regional Climate change (Kang et al. 2010) but there is a variability and lack of primary data on glaciers and Snow cover especially in tropical region. Himalayan cryosphere data can only be collected in a limited way on account of the remoteness, vastness, variability and inaccessibility of its terrain. In the present work, we present status of the Himalayan cryosphere and critically analyze this data.

2 Current Status

The Hindu Kush, Karakoram and Himalaya (HKKH) mountains along with Tibetan plateau are also the water tower of Asia and support population of more than a billion people (Immerzeel et al. 2010; Bookhagen and Burbank 2010). These regions have significantly large cryosphere area but the available areal extent of cryosphere is highly variable. The Himalayan-Karakoram ice cover varies from 40,775 km^2 (22,829 km^2 in the Himalaya and 17,946 km^2 in the Karakoram; Bolch et al. 2012) to ~53,178 km^2 (21,973 km^2 in the Himalaya and 21,205 km^2 in the Karakoram; Cogley et al. 2011 and Table 1). Estimates on the glacier volume is also highly variable and range from about 2300 km^3 (slope dependent ice thickness) to ~3600–~6500 km^3 based on volume area scaling (Bolch et al. 2012). According to Chinese Glacier Inventory (CGI), a number of 46,377 glaciers exist in

Table 1 Recent estimates of glacier area in (HKKH) Hindu-Kush-Karakoram Himalaya region

Himalaya (a)	Karakoram (b)	Total (a + b)	HKKH	References
35,109	–		–	Qin (1999)
33,050	16,600	49,650	–	Dyurgerov and Meier (1997)
33,050	15,400	48,450	–	Dyurgerov and Meier (2005)
–	–	–	1,14,800	WGMS (2008)
31,350	15,145	46,495		Raina (2009)
–	–	–	1,16,180	Xu et al. (2009)
21,973	21,205	53,178	–	Cogley et al. (2011)
–	–	–	60,000	Bajracharya et al. (2011), ICIMOD report
35,000	16,500	51,500	–	Barry and Gan (2011)
22,829	17,946	40,775	–	Bolch et al. (2012)
29,900	21,750	51,650	–	Kääb et al. (2012)
–	–	–	99,261	Yao et al. (2012)

Source Data is taken from given references

China covering an area of ~59,425 km^2 with total ice reserves of ~5600 km^3 (Li et al. 2008).

As per the mass balance studies by the Geological Survey of India (GSI), there are 9575 glaciers in the Indian Himalaya with a total glaciated area of 37,959 km^2. Raina (2009) has reported that the concentration of glaciers is highest in western Himalaya (Kashmir) and lowest in eastern Himalaya (Arunachal Pradesh) and that most of the glaciers are located in inaccessible terrain.

3 Climatic Zones

It is not appropriate to consider Hindu Kush-Karakoram-Himalaya (HKKH) as single domain because Hindu Kush-Karakoram mountains are separated from the eastern part of the Himalaya by ~2000 km. There is no gap between Himalaya and adjacent Hindu-Kush-Karakoram mountains but these regions differ significantly in terms of climatic (weather) conditions, particularly in source, and amount of precipitation (Armstrong 2010). The glaciers are strongly controlled by precipitation and these regions show uneven pattern of accumulation and ablation from east to west. Average annual precipitation varies from 1460 mm in the eastern Himalaya (Brahmaputra basin) to 413 mm in the western Himalayan (Indus basin) (Khan et al. submitted). A large amount of this precipitation falls in the form of snow at higher altitude domain due to cold temperature. Glaciers in the humid environment can exist at relatively lower elevations due to higher amount of snow accumulation and respond more sensitively to warming (Fujita and Nuimura 2011). The HKKH region is divided into 4 zones depending upon climatic conditions (UNEP 2012) (Fig. 1). The western most zone, marked as Zone 1, lies mainly in Afghanistan. The precipitation in this zone is dominated by westerlies. The glaciers in this zone are either relatively stable or retreating very slowly and are characterized by winter accumulation and summer ablation. This zone grades to Zone 2, which is represented by Karakoram and western Himalaya, where precipitation is dominated by westerlies. The summer monsoon also contributes to precipitation here, though its influence is limited. Glaciers in this zone are also characterized by 'winter accumulation' and 'summer ablation'. The glaciers, in this zone, exhibit advancing, stable and/or retreating nature but the rate of melting is relatively less intense and there are many surge glaciers with few large lakes. There is also spatial variability in glacier behaviour due to merging influence of westerlies and Indian summer monsoon. The central Himalaya encompassing mainly glaciers in India, southwestern Tibet and western Nepal fall in Zone 3, where glaciers are fed mainly by Indian monsoon and winter westerlies precipitation that starts from November and persist up to February/March. Relatively strong negative mass balance has been reported with in this zone (Kääb et al. 2012; Chaturvedi et al. 2014). The eastern most zone (i.e. Zone 4) includes mainly glaciers of eastern Nepal, Bhutan, Sikkim and southeastern Tibet. The precipitation in this zone is dominated by Indian summer monsoon. The glaciers in this zone are characterized by 'summer

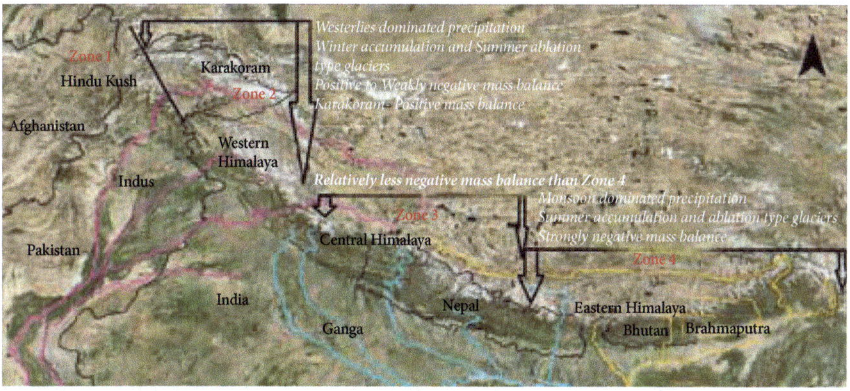

Fig. 1 Map showing 4 major different climatic zones along the Himalaya-Hindu-Kush-Karakoram regions based on UNEP (2012). *Zone 1* Lies mainly in Afghanistan and in this zone the precipitation is dominated by westerlies. The glaciers in this zone are either relatively stable or retreating very slowy and are characterized by winter accumulation and summer ablation. *Zone 2* Represented by Karakoram and western Himalaya, where precipitation is dominated by westerlies. The glaciers in this zone are advancing, stable and retreating but the rate of melting is relatively less intense and there are many India, southwestern Tibet and western Nepal. Relatively strong negative mass balance have been reported with in this. *Zone 4* The eastern most includes mainly eastern Nepal, Bhutan, Sikkim and southeastern Tibet and this zone is dominated by Indian summer monsoon. The glacier fragmentation, glacier thinning and growth of glacial lakes are significantly higher in this zone. *Source* Based on UNEP (2012)

accumulation and ablation type' with strongly negative mass balance (Kääb et al. 2012; Chaturvedi et al. 2014). The glacier fragmentation and thinning and growth of glacial lakes are significantly higher in this zone. The Indian summer monsoon results in an elevated heat pump here, causing higher glacier retreat (Fig. 1).

4 Mass Balance of Himalaya and the Adjacent Mountains

The available mass balance data on the Himalaya and the adjoining regions shows high variability. In the recent past, most of the world's mountain glaciers have been experiencing negative mass balance and terminal recession (Haeberli et al. 1999; Oerlemans and Grisogono 2002). Globally, glaciers have been showing a recession since last ice age (Barry 2006) and many Himalayan glaciers are retreating faster than the world average with thinning at the rate of 0.3–1 m/year (Dyurgerov and Meier 2005). A study using historical and other records since mid-nineteenth century showed that many small glaciers with size less than 0.2 km^2 have disappeared and the glaciers with relatively bigger size showed terminal recession at rates varying between 10 and 60 m/year (Bhambri and Bolch 2009). Gravimetric studies by GRACE satellite in high mountains of Asia during 2003–2009 suggest the average ice loss rate of 47 ± 12 Gt year^{-1}, equivalent to ~0.13 ± 0.04 mm year^{-1} sea level rise, which is twice as fast as the average rate over 4 decades (Matsuo and Heki 2010). They inferred that mass loss in Himalaya is slightly decelerating while those in northwestern glaciers show clear acceleration.

Studies in cryospheric domain of China have shown that the glaciated area in China has diminished by about 21% since the Little Ice Age. In the western part of China about 82.2% of glaciers have retreated during the last half century (Shi 2008; Liu et al. 2006). The retreat rate of glaciers on the Tibetan plateau has increased (Ren et al. 2004). Estimates show that since 1960 the glacial area has decreased by 4.5–7% (CNCC 2007), but the glaciers are relatively stable in the interior part of this region (Li et al. 1998, 2008). A study of 52 representative alpine glaciers in the eastern Hindu Kush range of Afghanistan and Pakistan for a period of about 31 years from 1976 to 2007, based on LANDSAT and ASTER images, showed that 76% of the studied glaciers were retreating, 16% were advancing while remaining 8% had stable terminus (Sarikaya et al. 2012). The authors also reported that the percentage of retreating glaciers increased from 41% (after 1992) to 76% (after 2001). There is a general trend of glacial melting not only in Himalaya but in many mountains glaciers of the world though the rate of melting varies from region to region (Armstrong 2010). On one side, while the rate of melting, in general has increased (almost twice) in recent time (Bajracharya et al. 2007) on the other hand, positive mass balance have been reported in Karakoram mountains (Scherler et al. 2011; Gardelle et al. 2012). Various studies also suggest that the Karakoram glaciers behave differently from the rest of the Himalaya and do not follow the global trend of retreat and melting (Zemp et al. 2009; Gardelle et al. 2012; Scherler et al. 2011; etc.). Recent history of Karakoram suggest that since the early 20th century, the glaciers in the Karakoram have diminished by about 5% or more whereas this decline was lower in the 1970s (Mayewski and Jeschke 1979) with slight advancement of few glaciers. Between the mid 1980 and 1990 glaciers again displayed a retreat with relatively minor loss in ice mass and thereafter started expanding in the high Karakoram Hewitt (2005, 2011).

Variable response of Himalayan glaciers to climate change is reported by Scherler et al. (2011). They identified six different geographic domains that differ in climate and topography namely, Hindu Kush, Karakoram, western Himalaya, West Kunlun Shah, North central Himalaya and South central Himalaya. These regions are characterized by the decreasing effect of monsoon and increasing influence of westerlies from east to west. These studies were based on analyses of 286 mountain glaciers by measuring frontal changes and surface velocities of the glaciers (excluding the surging glaciers) using remotely sensed ASTER and SPOT satellite images from the year 2000–2008. The role and influence of debris cover in the rate of glacial retreat was highlighted. Scherler et al. (2011) also reported that more than 65% of the glaciers that are influenced by monsoon are retreating whereas the glaciers which have thick debris cover with low gradient terminus have stable fronts. More than 50% of the glaciers of the Karakoram in the north-western part of Himalaya, which are influenced by westerlies, are either advancing or are stable. Another satellite based study reported positive mass balance (0.11 ± 0.22 m/year) for the Karakoram glaciers over the period 1999–2008 and the contribution of Karakoram glaciers to sea level at −0.01 mm/year (Gardelle et al. 2012). Their work contradicts the work of Scherler et al. (2011) which observes that the retreat rate of debris covered and debris free glaciers are similar. GRACE satellite gravimetric study (Matsuo and Heki 2010) stated that significant amount of glacial ice melted in the High mountains of Asia during the period of 2003–2009 with the average rate of 47 ± 12 Gt/a, which is equivalent to ~0.13 mm/year rise in sea level. Subsequently, another work (Jacob et al. 2012) based on simultaneous inversion of monthly GRACE-derived satellite gravity fields inferred a mass loss of only 4 ± 20 Gt/a in the high mountains of Asia during 2003–2010 which is significantly lower than the previously reported mass loss. Another mass balance study for five years period from 2003 to 2008 in the entire Hindu Kush-Karakoram-Himalayan region by Kääb et al. 2012 stated that the regional average glacier ice thinning rates are similar for debris covered and debris free glaciers. They reported a significantly more negative mass change of −12.8 ± 3.5 Gt/a and sea level rise of 0.035 ± 0.009 mm/year and a specific mass balance of 0.21 ± 0.05 m/year. The monitoring of seasonal snow cover in Hindu-Kush-Himalaya region using MODIS data from 2000 to 2010 show peak snow cover depletion during winter (Gurung et al. 2011). Seasonal snow cover was monitored in 28 river sub basins using normalized snow index (NDSI) method in central and western Himalaya by Kulkarni et al. (2011). Their study has shown a significant amount of snow melt even in the winter from October to December.

In the Nepal Himalaya, glaciers were first mapped by ICIMOD, 2001 from the published topographic maps by Survey of India. A total of 3808 glaciers with a total area of 3902 km^2 and 312 km^3 estimated ice reserves were reported and studied by using satellite images (Bajracharya et al. 2011.) Their study estimates the glacier area loss at about 24% and the loss of ice reserves at 29% between 1977 and 2010. As compare to earlier inventory of 2001, there has been 11% increase in the number of glaciers due to fragmentation of ice and shrinkage. The retreat rate for a few glaciers of Nepal Himalaya, for a longer span of time, is reported by Shrestha (2005). As per this study, the average rate of retreat of Rikha-Samba glacier in western Nepal for the period from 1974–1994, 1994–1998 and 1998–1999 are 10.8, 18.2 and 11.5 m/year respectively while that for Yala glacier in Central Nepal was 0.5 m/year from 1982–1987 and 3.86 m/year from 1987–1996. Another glacier, AX 010 in the central Himalaya has retreated with different rates in different periods. Between 1978 and 1989 it recorded retreat at 2.7 m/year while the rate of retreat increased to 14.0 m/year during 1989–1991. The retreat rate decreased to 3.0 m/year from 1991 to 1995 and relatively higher retreat onwards of ~20 m/year that reached up to 50 m/year during 1998–1999 (Shrestha 2005).

Bolch et al. (2012) studied mass changes for ten glaciers located south and west of Mt. Everest, Nepal over eight years period (1962–1970) using stereo Corona spy images, satellite imageries and Cartosat-1 data. They reported specific mass loss during the study period at 0.32 ± 0.08 m w.e.a^{-1}. A recent mass balance study (Chaturvedi et al. 2014) in the Himalayan-Karakoram region for the year 2000 (based on CMIP5 multi-model climate projection) reported glacial mass loss to be 6.6 ± 1 Gt/a, which is almost half that of earlier reports (Kääb et al. 2012). Their works shows significantly high variability in the mass balance along the Himalaya from east to west and Karakoram. In the western Himalaya, they reported a mass balance of −165 ± 8 mm w.e.a^{-1}, for the central Himalaya −508 ± 69 mm w.e.a^{-1} and in the eastern Himalaya −809 ± 38 mm w.e.a^{-1}. However for the Karakoram region significantly high positive glacial mass balance is reported at 171 ± 63 mm w.e.a^{-1}. Another work published at nearly the same time (Gardelle et al. 2013) contradict this (Chaturvedi et al. 2014) and reports contrasting pattern of glacier mass balance for 9 study sites between Pamir and Hengduan Shen (Eastern Himalaya). Their study shows that there is relatively less mass loss in the eastern and central Himalaya (−220 ± 120 mm w.e. $year^{-1}$ to −330 ± 140 mm w.e. $year^{-1}$) than western Himalaya (−450 ± 130 mm w.e. $year^{-1}$). In contrast slight positive mass balance is reported for the central Karakoram (100 mm w.e. $year^{-1}$) and in the western Pamir (140 mm w.e. $year^{-1}$).

The glacier inventory based on satellite data in the Bhutan Himalaya reported 885 glaciers in this region with a total ~642 ± 16.1 km^2 area which contributes ~1.6% of the total land area in Bhutan. The inventory results show 23.3 ± 0.9% loss in glacial area between 1980 and 2010. The ice loss was greater for debris free glaciers. The ice loss was highest (11.6 ± 1.2%) between 1980 and 1990 and lowest (6.7 ± 0.1%) between 2000 and 2010 (Bajracharya et al. 2014a, b).

Kulkarni et al. (2011) have monitored the Indian Himalaya by using remote sensed data. They have discussed the influence of climate change on glaciers, seasonal snow cover and moraine-dammed lakes. Their studies reveal an overall reduction in glacier area ~16% from 6332 to 5329 km^2 since 1962 for 1868 glaciers in the Chenab, Parbati and Baspa basins where as the number of glaciers increased due to fragmentation. Seasonal snow cover monitoring in the 28 river sub basins of the central and western Himalaya by using normalized difference snow index (NDSI) suggest a large amount of snow retreat during the early part of winter even in higher altitude region (3000 m). Earlier studies in the same Indian domain i.e. Chenab, Parbati and Baspa basins by Kulkarni et al. 2007 with the aid of Indian remote sensing satellite data from 1962 to 2001–2004 have been attempted. In their study of 466 glaciers, they report an area loss from 2077 to 1628 km^2, an overall deglaciation of 21%. In another study using optical sensor data of the Chenab basin, Shukla et al. (2009) suggest that the glacier area has reduced from a total of 110.5–96.8 km^2 with an overall deglaciation of 13.7 km^2. They have also studied effect of debris cover on the glacier and its temporal variation in the Chenab basin by using optical satellite sensor has been studied. They infer that the glacier snout receded about 756 m and the total glaciated area has reduced by 13.7 km^2 from 1963 to 2004. Recently, Bahuguna et al. (2014) found the change in the extent of Himalayan glaciers during last decade by using IRS LISS III images. They monitored 2018 glaciers from different climatic zones along the Himalaya (East and West Himalaya) and Karakoram, and reported that among these, 1752 glaciers (86.8%) have stable fronts, 248 glaciers (12.3%) have retreated and 18 glaciers

(0.9%) show snout advancement. The net loss of the studied glaciers was found to be 20.94 km^2.

High altitude-small glaciers in the Ladakh region displayed a decrease in glaciated area of about 14% (0.3%/year) with average terminus retreat of about 3 m/year during 1969–2010 (Schmidt and Nüsser 2012). In Spiti/Lahaul region of Himachal Pradesh, western Himalaya the glaciers experienced rapid ice loss (−0.7 to −0.8 m w.e./year) between 1999 and 2004 (Berthier et al. 2007), which is approximately two times higher than the average mass balance between 1977 and 1999 (−0.34 m w.e. $year^{-1}$) for the Himalaya (Dyurgerov and Meier 2005). The behavior of Chhota Shigri glacier in this region has been studied by Azam et al. (2012). This glacier experienced a mass loss of −0.67 m w.e. $year^{-1}$ from 2002 to 2010. Another study on this glacier by Vincent et al. (2013) indicates only a slight mass loss (−3.8 ± 2.0 m w.e.) which corresponds to 0.17 ± 0.09 m w.e. $year^{-1}$ between 1988 and 2010. They reported negative mass balance (−4.8 ± 1.8 m w.e.) between 1999 and 2010 corresponding to −0.44 ± 0.16 m w.e $year^{-1}$ and a slight positive or near Zero mass balance between 1988 and 1999. Earlier, Kulkarni et al. (2011) reported that the snow line at the end of ablation season on the Chhota Shigri glacier has changed from 4900 to 5200 m a.s.l. from the late 1970 to present. A relatively small, Hamtah glacier in the adjoining area has been studied in detailed by Geological survey of India. The average mass balance here was found to be $-4.757 \times 106\ m^3$ of w.e. with an average rate of ice loss 1.45 m and the loss in ice thickness from 2000 to 2009 of 13.11 m. Samundra Tapu glacier which is drained in Chandra basin in Himachal Pradesh has been studied in detailed by Kulkarni et al. (2006). They reported that total recession of glacier during a period of 38 years from 1962 to 2000 was about 19.5 m/year. The glacier extent was significantly reduced 73–65 km^2 during this period with overall 11% deglaciation. Average recession rates for some glaciers in the Himalaya, as reported by Shrestha, (2005) are given in Table 2.

Numerous studies have estimated the retreat and mass balance for short and long period of individual glaciers in the Himalayan domain. Gangotri glacier which lies in the Uttarakhand Himalaya and drained in the Bhagirathi basin has been retreating three

Table 2 Average rate of retreat of selected glaciers Shrestha (2005)

Glacier	Period	Average rate of retreat in m/year
Gangotri, Uttarakhand	1935–1976	14.6
Do	1977–1990	28
Pindari, Uttarakhand	1845–1966	135.2
Milam, Uttarakhand	1849–1957	12.5
Do	1909–1984	13.2
Ponting, Uttarakhand	1906–1957	5.1
Triloknath, Himachal Pradesh	1969–1995	15.4
Chota Shigri, Himachal Pradesh	1986–1995	6.7
Bada Shigri, Himachal Pradesh	1977–1995	36.1
Zemu glacier, Sikkim	1909–1965	0.2
Do	1977–1984	27.7

Source Shrestha (2005)

times faster over the last 30 years than during the preceding 200 years (Srivastava 2012). The results show that Gangotri glacier has retreated by 1147 m and the total area vacated during 1935–1996 is estimated to be 0.58 km^2. A study of data for about 4 decades shows that this glacier is receding on an average rate of about 18–19 m/year (Bhambri et al. 2011). Dokriani glacier, which also lies in the Bhagirathi basin receded by 550 m from 1962 to 1995 with an average rate of 16.6 m/year (Dobhal et al. 2004). During this period the glacier volume has reduced by about 20% and frontal area have reduced by about 10%. The snout of the Dokriani glacier has vacated an area of 3957 m^2 from 1991 to 1995 with average rate of recession of 17.4 m/year. The Bhagirathi-Kharak glacier and the adjacent Satopant glacier has been studied by Nainwal et al. 2008 by the help of Survey of India toposheets and Total Station Survey from 1962 to 2006. The total area vacated by the Satopant and Bhagirathi Kharak glaciers during 44 years was 0.314 and 0.13 km^2 respectively. They reported that the average annual retreat of the snouts of the Satopant and Bhagirathi Kharak glaciers were $\sim$23 m/year and $\sim$7 m/year respectively. Bhambri et al. 2011 reported that Satopant glacier lost 0.280 km^2 and Bhagirathi Kharak glacier lost 0.083 km^2 from 1968 to 2006. As per their study, the Gangotri glacier shrank by 4.4 $\pm$ 2.7 km^2 with average rate of 0.11 $\pm$ 0.07 km^2 per year between 1968 and 2006.

The glacial mass balance studies, basin and sub-basin wise, also shows high variability. Over a period of twenty years (1990–2010) mass balance study in the Tista basin of Sikkim in eastern Himalaya suggest that the glaciers have lost an area of 3.3 $\pm$ 0.8% during the study period (Basnett et al. 2013). They also observed that debris mantled glaciers with lakes loose a greater area than debris covered glaciers not associated with lakes and the clean glaciers. In the Chandra Bhaga basin, north western Himalaya over a period of 30 years, the glaciated area has reduced from 377.6 km^2 in 1980 to 368.2 km^2 in 2010 suggesting a loss of 2.5% (Pandey and Venkataraman 2013). They estimated the glacier snout changes by satellite imageries and reported that the glacier snout have retreated 65.5 $\pm$ 169.1 m during three decades with an average rate of 15.5 $\pm$ 5.6 m/year.

In the Garhwal Himalyan region, a study using remote sensed corona and ASTER satellite images from 1968 to 2006 has been carried out by Bhambri et al. (2011). As per them, the loss of glaciated area in the upper Bhagirathi basin was 9.0 $\pm$ 7.7 km^2 (3.3 $\pm$ 2.8%) while in upper Saraswati/Alaknanda basins it was 18.4 $\pm$ 9.0 km (5.7 $\pm$ 2.7%), which is almost double than that in the Bhagirathi basin. This difference in ice loss can be attributed mainly to the difference in glacier size. Debris covered glacier area also increased in the Saraswati/Alaknanda (17.8 $\pm$ 3.1%) basin as compared to upper Bhagirathi basin (11.8 $\pm$ 3.0%) from 1968 to 2006. They also reported that due to fragmentation of ice bodies, the number of glaciers increased from 69 to 75 during the period from 1968 to 2006. Surface area loss by $\sim$18% and retreat of snout at $\sim$535 m with an average rate of 13.4 m a^{-1} has been estimated for Tipra glacier (draining in the Alaknanda basin), from 1962 to 2002. The major part of Tipra glacier has detached from the main trunk and separated into Tipra and Ratan glaciers after 1987 (Mehta et al. 2011). They report annual retreat rate for Tipra glacier at 21.3 and 21.2 m a^{-1} for Ratan glacier during 2002–2008.

5 Glacial Melt Fraction

The rivers originating from the higher Himalaya and the adjacent mountains are among the most melt-water-dependent on this planet (Lutz et al. 2014) and feed the three major river system—The Indus, the Ganga and the Brahmaputra which support a large population in the downstream areas. Snow and glacier melt contributions to the Himalayan river discharge are either unknown or poorly known. Some previous studies on a regional scale show high variability (Immerzeel et al. 2010; Immerzeel and Bierkens 2012; Jianchu et al. 2009). In the recent time, however few site specific studies have either overestimated the glacial melt fraction or are doubtful (Schaner et al. 2012; Khan et al. 2017). Studies show that the water flow in the Himalayan rivers is mostly driven by precipitation and that the glacier melt is only a minor contributor to the total river discharge (Immerzeel et al. 2010).

Water is stored in glaciers and ice sheets/ice caps for longer period of time, from years to centuries and on seasonal scale as snow and ice from months to years. Long period storage affects global sea level and water balance of glaciated catchments and is very important for water resources especially in arid semi-arid regions (Jansson et al. 2003). Lot of studies related to mass balance, described above, show negative mass balance which means glacier melt is contributing to river discharge but unfortunately we do not have any reliable data on the glacial melt fraction.

The precipitation in the solid form particularly in the mountainous regions can be stored for a longer period of time as perennial snow and ice or for a shorter period as seasonal snow before melting. More than half of the world's water supply comes from the global river runoff and more than one-sixth of the earth's population depends on glaciers and snow packs for the water supply (Barnett et al. 2005). The Himalaya and the adjacent mountains are the water towers of the Asia. These mountains act as orographic barriers characterized by high precipitation and contribute a significant amount of snow and ice melt to the river discharge (Messerli 2000; Bookhagen and Burbank 2010). The precipitation pattern in the Himalaya is not same and it varies from east to west. The Indian summer monsoon precipitation dominates in the eastern Himalaya whereas westerly and cyclonic storms dominate in the western Himalaya which contributes about two-third of annual precipitation (mainly in the form of snow). The source of remaining precipitation (one-third) is the summer monsoon (Wake 1989). As a consequence of this precipitation pattern, basin runoff decreases from the east to west direction (Table 3). Due to this climatic difference along the Himalayan region, Rees and Collins (2006) suggested that if the entire Himalaya and the adjacent mountain were to disappear, there would be a significant reduction in annual flow of rivers (about 33% in the west and only 4–18% in the east).

A study of nine sub-basins in the Nepal Himalaya by Alford and Armstrong (2010) suggested that the glacial melt contribution to sub-basin stream flow varies significantly from ~2% in the Likhu Khola basin to ~20% in the Budhi Gandaki basin. The average across nine basins in Nepal was ~10%. They also reported that the total ice volume of Nepal Himalaya, which is ~480 km^3, represented only

Table 3 Compilation of cryosphere and hydrosphere

Basin	Total basin area (km^2)	Glaciated area in km^2	Glaciated area (%)	Number of glaciers	Glacier melt (%)	Ice reserves (km^3)	Annual basin precipitation in (mm)	Mean discharge (m^3/s)	References
Indus	–	8,926	–	5,057	–	850	–		(1)
	1,081,718	–	–	–	44.8	–	–	5,533	(2)
	1,005,786	–	2.2	–	26	–	423	–	(3)
	1,139,814	20,325	1.78	–	–	–	405	–	(4)
	3,21,289+	–	–	–	–	–	1,097	–	(5)
	1,18,0000	32,246	–	16,049	–	–	–	–	(6)
	–	21,193	–	18,495	–	2,696	–	–	(7)
	–	–	–	–	–	–	413	–	(8)
Ganga	–	16,677	–	6,694		1,971	–	–	(1)
	10,16,124	–	–	–	9.1	–	–	18,691	(2)
	9,90,316	–	1.0	–	3.1	–	1,035	–	(3)
	10,23609	12,659	1.24	–	–	–	1,118	–	(4)
	8,61,452+		–	–	–	–	1,051	–	(5)
	9,60,000	18,393	–	6,237	–	–	–		(6)
	–	9,012	–	7,963	–	793.5	–	–	(7)
	–	–	–	–	–	–	1,081	–	(8)

(continued)

Table 3 (continued)

Basin	Total basin area (km^2)	Glaciated area in km^2	Glaciated area (%)	Number of glaciers	Glacier melt (%)	Ice reserves (km^3)	Annual basin precipitation in (mm)	Mean discharge (m^3/s)	References
Brahmaputra	–	6,579	–	4,366	–	600	–	–	(1)
	19,824	–	–	–	12.3	–	–	19,824	(2)
	5,25,797	–	3.1	–	<10	–	1,071		(3)
	5,27,666	16,118	3.05	–	–	–	1,145	–	(4)
	1,94413+	–	–	–	–	–	2,589	–	(5)
	6,30,000	20,543	–	10,106	–	–	–	–	(6)
	–	14,020	1.12	11,497	–	1,303	–	–	(7)
	–	–		–	–		1,460	–	(8)

Variability of the Indus basin, the Ganga and Brahmaputra basins

Note (1) Qin (1999), (2) Jianchu et al. (2009), (3) Immerzeel et al. (2010) and Immerzeel and Bierkens (2012), (4) Kaser et al. (2010), (5) Jain and Kumar (2012), (6) Sharma et al. (2013), (7) Bajracharya et al. (2014a, b), (8) Khan et al. Submitted

~4% of the total annual flow in the Ganga basin. Subsequently, another study in Nepal Himalaya, employing remote sensing and isotopic techniques, estimate glacial melt in the Langtang basin, only at 10% of total discharge (Racoviteanu 2011). The contribution of annual glacier melt to stream flow in the Trishuli and Dudh Kosi basins by using elevation dependent ice ablation model based on glacial areas from ASTER and IKONOS remote sensed data combined with hypsometry from SRTM have been estimated by (Racoviteanu et al. 2013). They inferred that glacial ice melt was positively correlated with the glaciated basin area and contributed 58.3% to the annual flow in the small Langtang Khola watershed in the Trishuli basin (43.5% glaciated area) and 21.2% in the Hinku watershed in Dudh Kosi basin (34.7% glaciated area). The percentage of glacial melt decreases significantly with downstream locations and becomes only 4.5% about 75 km away from the glacier snout.

The three river basins of the northern part of India i.e. The Indus, the Ganga and the Brahmaputra basins contribute more than half of the river discharge of India but the importance of glacial melt runoff varies significantly within these basins. The Indus river originates from the Tibetan plateau near Lake Mansarovar and Mount Kailash and after traveling a distance of ~2880 km enters the Arabian Sea. The Indus basin covers four countries and the major part of the basin lies in Pakistan (47%), followed by India (39%) with subordinate part in China (8%) and Afghanistan (6%), (Shrestha et al. 2015). The Ganga river originates from the snout of Gangotri glacier (Gaumukh) at an altitude of ~4000 m a.s.l. in the Uttarkashi district of Uttarakhand, central Himalaya, India. The total length of the Ganga river, from it source up to its fall in the Bay of Bengal, is ~2525 km. The Ganga basin covers four countries and the major part of the basin lies in India (79%), subordinate part in Nepal (14%) and only minor part in China (3%) and Bangladesh (4%), (Shrestha et al. 2015).

The river Brahmaputra originates from Angsi glacier in Mount Kailash, south Tibet at an elevation of 5300 m a.s.l. and enters into the Bay of Bengal after traveling a distance of ~2900 km. The Brahmaputra basin covers four countries. The major part of the basin falls in China (50%), followed by India (36%) and only minor part in Bhutan (7%) and Bangladesh (7%), (Shrestha et al. 2015). It has been estimated that ~75% of the Ganga basin is below 500 m a.s.l. compared with about ~25% for the Brahmaputra basin (Seidel et al. 2000).

There is wide variability in the published data for these three basins including drainage basin area, Glaciated area, number of glaciers drained, annual precipitation and glacial melt fraction which is shown in Table 3. By using Normalized melt index model in the Higher Himalayan regions, Immerzeel et al. (2010) stated that the glacial melt is very important for the Indus basin, important for the Brahmaputra basin and relatively less important for the Ganga basin.

Run off due to glacial melt is highest in the drier, westerly dominated Indus basin and relatively low in the wetter monsoon influenced catchments of Ganga and the Brahmaputra basins. The available estimates on the glacial melt estimation are highly variable: For Indus basin, it varies from 44.8 to 26%, in the Ganga basin it varies from 9.1 to 3% and for the Brhamaputra basin the glacial melt fraction varies from <10 to 12% (Immerzeel et al. 2010; Jianchu et al. 2009). (Immerzeel and Bierkens 2012; Jianchu et al. 2009). In the Upper Indus basin, the glacier and snow melt contribution is ~63% with minor contribution from rainfall (~27%). In the Upper Ganga and Brahmaputra basin the glaciers melt contribution in the stream flow is only ~20 and ~25% respectively. The major contribution is from rainfall with ~66% in the Upper Ganga basin and ~59% in the upper Brahmaputra basin (Lutz and Immerzeel 2013). A recent study by using cryosphere-hydrological modelling, suggests that the stream flow is dominated by glacial melt contributing to 40.6% in the Upper Indus Basin and 11.5% in the upper Ganga basin (Lutz et al. 2014). Earlier (Seidel et al. 2000) used Snow melt run off model and concluded that the snow melt contributes about 27% in the Brahmaputra basin and only 9% in the Ganga basin to the total runoff. In the western Himalaya catchment (Indus and Sutlej), snowmelt contributes ~50% in the annual runoff budget whereas in the central and eastern Himalaya (except the large Tsangpo/Brahmaputra catchment), the snow melt contribution is less than 20% in the annual runoff budget. The snow melt contribution is slightly higher in the Tsangpo/Brahamaputra basin and reaches up to ~34% (Bookhagen and Burbank 2010). The average snow and glacier melt contribution in the Satluj river flow at Bhakra Dam, by using Water Balance approach for a period of 10 years, was found to be 59% and the remaining 41% is from rainfall (Singh and Jain 2002). In Chenab river basin at Akhnoor (Singh et al. 1997) estimated 49% snow and glacier melt contribution. In Beas river at Pandoh Dam (Kumar et al. 2007) reported 35% snow and glacier melt contribution. Singh et al. (2006) simulate the melt runoff from a highly glacierized small basin for the summer season in the Dokriani melt stream in the central Himalayan region. Based on a 2-year simulation, it is found that, on average, the contributions of glacier melt in the total runoff is 87% and the remaining 13% is from rainfall. Arora et al. (2010) estimated the melt contribution in upper Bhagirathi and Alaknanda basins by using SNOWMOD model. They reported that the average contribution of the snowmelt, rainfall runoff and base flow in the Bhagirathi river at Loharinag project site during 1999–2002 were 70.54, 16.30 and 13.14% respectively. The average contribution of snowmelt, rainfall runoff and base flow in the Alaknanda river at Tapovan Vishnugad project site during 1983–1987 were 77.35, 10.35 and 12.32% respectively. It is also reported that the snow and glacier melt contribution to the total runoff in the Bhagirathi basin at Devprayag is 28.7% and the remaining 71.3% contribution is from rainfall and base flow (Singh et al. 2010). By using SNOWMOD model, Singh et al. (2008) reported that the average glacier melt contribution in the Bhagirathi river at Bhojwasa (close to Gaumukh) is 97% and the rainfall contributes only 3%. Recently, Schanner et al. 2012 used energy balance approach and reported that the glacial melt fraction in the upper Ganga basin is only 6.5%. With the help of three component-end member mixing model, by using stable oxygen isotopes and electrical conductivity, the glacial melt fraction in the upper Ganga river at Rishikesh was reported at ~32% (and maximum up to 40%) during the monsoon season (Maurya et al. 2011). Recently, Khan et al. (2017) have used the same end member mixing model and report that the glacial melt fraction in the Upper Ganga-Bhagirathi basin at Devprayag is only 11% during pre-monsoon and

12% during post-monsoon periods. The influence of glacial melt fraction in the river decreases as we move away from the source (i.e. from the upstream to downstream).

6 Discussion

Though several glacier inventories have been published for HKKH region or for some specific part of these regions, in particular for the Himalayan regions (Cogley et al. 2011; Bolch et al. 2012; Kääb et al. 2012; etc.) but their estimates of extent and volume of Himalayan cryosphere are highly variable (Table 3). Armstrong (2010) suggested that a long term consistent glacier database is needed for better understanding of climate change impacts on glaciers and also for hydrological monitoring in the Himalaya.

HKKH region is highly inaccessible, has rough terrain, highly variable climatic zones across east to west and in the adjacent mountainous regions which complicates understanding of the glacier behaviour. The seasonal snow cover, which has relatively higher residence period (up to six months) along with the debris cover make it further difficult to delineate the number and aerial extent of glaciers (Bolch et al. 2012). The hydrological budget of the higher Himalayan rivers depends on the precipitation in the form of snowfall at higher altitude and in the form of rainfall in the lower altitude regions. The climate of the HKKH region is not homogenous and is fed dominantly by different moisture sources. The precipitation in the eastern and central Himalaya is dominated by Indian summer monsoon which brings moisture from the Indian Ocean and the Bay of Bengal and strikes along the Himalayan mountains, which act an orographic barrier. The influence of monsoon decreases towards the western side of this region. The western Himalaya, Karakoram and Hindu Kush regions are dominated by westerlies which bring moisture from the Mediterranean and Caspian seas and nourish this zone. The glaciers in these regions are characterized by winter accumulation and summer ablation. The health of glaciers is strongly dependent on the precipitation but the available estimates on precipitation are highly variable and in few cases it is either unacceptably high or unacceptably low. The basin wise precipitation variability is shown in Table 3. For the Indus basin the variability is more than 250%, for the Ganga basin it is 100% and for the Brahmaputra basin the variability is more than 240%. This problem is addressed by a critical analysis, optimum use of remote sensed moisture data and validation with limited ground data to arrive at reliable annual precipitations for the three basins (Khan et al. submitted). The glacial melt contribution in different climatic zones varies significantly but overall the contribution from glacier melts to the downstream river discharge is relatively small in the three Himalayan basins.

From the above review, it is clear that the reported available data on almost every component of Himalayan cryosphere and adjacent mountains shows high variability. On account of its vast ice area and inaccessibility, the observations/ measurements in respect of different parameters need to be carried out in a well organized way. Limited site specific studies of some glaciers in a particular domain will not represent the characteristic for other glaciers in different climatic zones. The only strong option to solve this variability appears to be to adopt the cryosphere modelling.

The available estimates on the glacial melt fraction also show high variability, for example for the Indus basin the variability is ~170%, for the Ganga basin it is ~300% and for the Brahmaputra basin the variability is more than 100%. The number of glaciers in the Himalaya and the adjacent mountains are only roughly known with high uncertainty. The reported basin wise glaciated areas also varies from 300% for Indus basin, 200% for the Ganga basin and it is more than 300% for the Brahmaputra basin.

Thus, it is well demonstrated that despite significance of cryosphere for the HKKH hydrosphere, not only the past behaviour of this component is poorly understood but even the present day data on cryosphere is not well known. This is despite the well accepted understanding that this represents the one component which is likely to be the first to respond to the patterns of complex climate changes (Armstrong 2010). Besides the basic issues of quantifying the cryosphere, there are other aspects such as snow melt versus glacier run off contribution to the Himalayan rivers, long-term measurement series of multiple parameters at high altitude and reliable quantification of mass balance which require immediate attention.

Acknowledgements This work was supported by Science and Engineering Research Board, Department of Science and Technology funded project entitled "Fingerprinting of glacial melt water in the Ganga basin-Implications for modelling of hydrological cycle in a Himalayan river system" grant number NO.SR/DGH-46/2012.

References

Alford D, R. Armstrong (2010) The role of glaciers in stream flow from the Nepal Himalaya. Cryosphere Discuss 4(2):469–494

Armstrong RL (2010) The Glaciers of the Hindu-Kush-Himalayan Region: a summary of the science regarding glacier melt/retreat in the Himalayan, Hindu Kush, Karakoram, Pamir, and Tien Shan Mountain Ranges. Technical Paper. International Centre for Integrated Mountain Development (ICIMOD). Kathmandu, Nepal

Arora M, Rathore DS, Kumar R, Kumar A (2010) Estimation of melt contribution to total streamflow in River Bhagirathi and River DhauliGanga at Loharinag Pala and TapovanVishnugad project sites. J Water Resour Protect 2(7):636–643. doi:10.4236/jwarp.2010.27073

Azam MF, Wagnon P, Ramanathan A, Vincent C, Sharma P, Arnaud Y, Linda A, Pottakkal JG, Chevallier P, Singh VB, Berthier E (2012) From balance to imbalance: a shift in the dynamic behaviour of Chhota Shigri glacier, western Himalaya, India. J Glaciology 58(208):315–324

Bahuguna IM, Rathore BP, Brahmbhatt R, Sharma M, Dhar S, Randhawa SS, Kumar K, Romshoo S, Shah RD, Ganjoo RK (2014) Are the Himalayan glaciers retreating. Curr Sci 106 (7):1008–1013

Bajracharya SR, Maharjan SB, Shrestha F (2014a) The status and decadal change of glaciers in Bhutan from the 1980s to 2010 based on satellite data. Ann Glaciol 55(66):159–166

Bajracharya SR, Maharjan SB, Shrestha F (2014b) Understanding dynamics of Himalayan glaciers: scope and challenges of remote sensing. Int Arch Photogrammetry Remote Sens Spat Inf Sci 40(8):1283

Bajracharya SR, Mool PK, Shrestha BR (2007) Impact of climate change on Himalayan glaciers and glacial lakes: Case studies on GLOF and associated hazards in Nepal and Bhutan. Kathmandu, Nepal, International Centre for Integrated Mountain Development (ICIMOD Report)

Bajracharya SR, Shrestha F, Maharjan SB (2011) Glaciers shrinking in Nepal Himalaya. INTECH Open Access Publisher

Barry RG (1987) The Cryosphere-neglected component of the climate system. In: Radok U (ed) Towards understanding climate. West view Press, Boulder, CO, pp 35–67

Barry RG (2002) History of the World Data Center for Glaciology, Boulder, and National Snow and Ice Data Center at the University of Colarado. Glacial, Data Report GD-30. NSDC, Co

Barry RG (2006) The status of research on glaciers and global glacier recession: a review. Prog Phys Geogr 30(3):285–306

Barry R, Gan TY (2011) The global cryosphere: past, present and future. Cambridge University Press, Cambridge

Barnett TP, Adam JC, Lettenmaier DP (2005) Potential impacts of a warming climate on water availability in snow-dominated regions. Nature 438(7066):303–309

Basnett S, Kulkarni AV, Bolch T (2013) The influence of debris cover and glacial lakes on the recession of glaciers in Sikkim Himalaya India. J Glaciol 59(218):1035–1046

Berthier E, Arnaud Y, Kumar R, Ahmad S, Wagnon P, Chevallier P (2007) Remote sensing estimates of glacier mass balances in the Himachal Pradesh (Western Himalaya, India). Remote Sens Environ 108(3):327–338

Bhambri R, Bolch T (2009) Glacier mapping: a review with special reference to the Indian Himalayas. Prog Phys Geogr 33(5):672–704

Bhambri R, Bolch T, Chaujar RK, Kulshreshtha SC (2011) Glacier changes in the Garhwal Himalaya, India, from 1968 to 2006 based on remote sensing. J Glaciol 57(203):543–556

Bookhagen B, Burbank DW (2010) Toward a complete Himalayan hydrological budget: spatiotemporal distribution of snowmelt and rainfall and their impact on river discharge. J Geophys Res 115(F03019):2009J. doi:10.1029/F001426

Bolch T, Kulkarni A, Kääb A, Huggel C, Paul F, Cogley JG, Frey H, Kargel JS, Fujita K, Scheel M, Bajracharya S, Stoffel M (2012) The state and fate of Himalayan glaciers. Science 336:310–314

Chaturvedi RK, Kulkarni A, Karyakarte Y, Joshi J, Bala G (2014) Glacial mass balance changes in the Karakoram and Himalaya based on CMIP5 multi-model climate projections. Clim Change 123(2):315–328

CNCC (2007) China national report on climate change, China National Committee on Climate Change. Beijing, China

Cogley JG, Hock R, Rasmussen LA, Arendt AA, Bauder A, Braithwaite RJ, Jansson P, Kaser G, Möller M, Nicholson L, Zemp M (2011) Glossary of glacier mass balance and related terms, IHP-VII technical documents in hydrology No. 86, IACS Contribution No. 2. International Hydrological Program, UNESCO, Paris

Dobhal DP, Gergan JT, Thayyen RJ (2004) Recession and morphogeometrical changes of Dokriani glacier (1962–1995), Garhwal Himalaya India. Curr Sci 86(5):692–696

Dobrowolski, A.B. 1923. Historjanaturalna lodu (Natural history of ice). *Warsaw:Nakland H. Lindenfela. 99–940* (in polish, French summary)

Dyhrenfurth GO (1955) To the third pole: the history of the high Himalaya, 1st edn. Ex Libris, Werner Laurie, London

Dyurgerov MB, Meier MF (1997) Mass balance of mountain and subpolar glaciers: a new global assessment for 1961–1990. Arctic Alp Res 379–391

Dyurgerov MB, Meier MF (2005) Glaciers and the changing earth system: a 2004 snapshot, vol 58. Institute of Arctic and Alpine Research, University of Colorado, Boulder

Fujita K, Nuimura T (2011) Spatially heterogeneous wastage of Himalayan glaciers. Proc Natl Acad Sci 108(34):14011–14014

Gardelle J, Berthier E, Arnaud Y (2012) Slight mass gain of Karakoram glaciers in the early twenty-first century. Nat Geosci 5:322–325

Gardelle J, Berthier E, Arnaud Y, Kääb A (2013) Region-wide glacier mass balances over the Pamir-Karakoram-Himalaya during 1999–2011 (vol 7, pp 1263, 2013). Cryosphere 7(6):1885–1886

Gurung DR, Kulkarni AV, Giriraj A, Aung KS, Shrestha B, Srinivasan J (2011) Changes in seasonal snow cover in Hindu Kush-Himalayan region. Cryosphere Discuss 5(2):755–777

Haeberli W, Frauenfelder R, Hoelzle M, Maisch M (1999) On rates and acceleration trends of global glacier mass changes. *GeografiskaAnnaler: Series A*. Phys Geogr 81(4):585–591

Hewitt K (2005) The Karakoram anomaly? Glacier expansion and the 'elevation effect', Karakoram Himalaya. Mt Res Dev 25(4):332–340

Hewitt K (2011) Glacier change, concentration, and elevation effects in the Karakoram Himalaya, Upper Indus Basin. Mt Res Dev 31(3):188–200

Immerzeel WW, Bierkens MFP (2012) Asia's water balance. Nat Geosci 5(12):841–842

Immerzeel WW, Van Beek LPH, Bierkens MFP (2010) Climate change will affect the Asian water towers. Science 328:1382–1385

Jacob T, Wahr J, Pfeffer WT, Swenson S (2012) Recent contributions of glaciers and ice caps to sea level rise. Nature 482(7386):514–518

Jain SK, Kumar V (2012) Trend analysis of rainfall and temperature data for India. Curr Sci 102(1):37–49

Jansson P, Hock R, Schneider T (2003) The concept of glacier storage: a review. J Hydrol 282 (1):116–129

Jianchu X, Shrestha A, Eriksson M (2009) Climate change and its impacts on glaciers and water resource management in the Himalayan Region. Assessment of Snow, Glaciers and Water Resources in Asia. International Hydrological Programme of UNESCO and Hydrology and Water Resources Programme of WMO. Koblenz, Germany, pp 44–54

Kääb A, Berthier E, Nuth C, Gardelle J, Arnaud Y (2012) Contrasting patterns of early twenty-first-century glacier mass change in the Himalayas. Nature 488(7412):495–498

Kaser G, Großhauser M, Marzeion B (2010) Contribution potential of glaciers to water availability in different climate regimes. Proc Natl Acad Sci 107(47):20223–20227

Kang S, Xu Y, You Q, Flügel WA, Pepin N, Yao T (2010) Review of climate and cryospheric change in the Tibetan Plateau. Environ Res Lett 5(1):015101

Khan AA, Pant NC, Sarkar A, Tandon SK, Thamban M, Mahalinganathan K (2017) The Himalayan cryosphere: a critical assessment and evaluation of glacial melt fraction in the Bhagirathi basin. Geosci Front 8(1):107–115

Khan AA, Pant NC, Ravindra R, Alok A, Gupta M, Gupta S. A precipitation perspective of the hydrosphere-cryosphere interaction in the Himalaya (submitted in geological society of London)

Kulkarni AV, Rathore BP, Singh SK, Bahuguna IM (2011) Understanding changes in the Himalayan cryosphere using remote sensing techniques. Int J Remote Sens 32(3):601–615

Kulkarni AV, Dhar S, Rathore BP, Kalia R (2006) Recession of samudra tapu glacier, chandra river basin, Himachal Pradesh. J Indian Soc Remote Sens 34(1):39–46

Kulkarni AV, Bahuguna IM, Rathore BP, Singh SK, Randhawa SS, Sood RK, Dhar S (2007) Glacial retreat in Himalaya using Indian Remote Sensing satellite data. Curr Sci 92(1):69–74

Kumar V, Singh P, Singh V (2007) Snow and glacier melt contribution in the Beas river at Pandoh dam, Himachal Pradesh India. Hydrol Sci J 52(2):376–388

Li Z, Sun W, Zeng Q (1998) Measurements of glacier variation in the Tibetan Plateau using Landsat data. Remote Sens Environ 63(3):258–264

Li X, Cheng G, Jin H, Kang E, Che T, Jin R, Wu L, Nan Z, Wang J, Shen Y (2008) Cryospheric change in China. Glob Planet Change 62(3):210–218

Liu SY, Ding YJ, Zhang Y, Shangguan DH, Li J, Han HD, Wang J, Xie CM (2006) Impact of the glacial change on water resources in the Tarim River Basin. Acta Geogr Sin 61(5):482–490

Lutz AF, Immerzeel WW (2013) Water availability analysis for the upper Indus, Ganges, Brahmaputra, Salween and Mekong river basins. Final Report to ICIMOD, September 2013. Future Water Report (127)
Lutz AF, Immerzeel WW, Shrestha AB, Bierkens MFP (2014) Consistent increase in High Asia's runoff due to increasing glacier melt and precipitation. Nat Clim Change 4(7):587–592
Matsuo K, Heki K (2010) Time-variable ice loss in Asian high mountains from satellite gravimetry. Earth Planet Sci Lett 290(1):30–36
Maurya AS, Shah M, Deshpande RD, Bhardwaj RM, Prasad A, Gupta SK (2011) Hydrograph separation and precipitation source identification using stable water isotopes and conductivity: River Ganga at Himalayan foothills. Hydrol Process 25(10):1521–1530
Mayewski PA, Jeschke PA (1979) Himalayan and Trans-Himalayan glacier fluctuations since AD 1812. Arctic and alpine research, pp 267–287
Mehta M, Dobhal DP, Bisht MPS (2011) Change of Tipra glacier in the Garhwal Himalaya, India, between 1962 and 2008. Prog Phys Geogr 35(6):721–738
Meier MF, Bahr DB (1996) Counting glaciers: use of scaling methods to estimate the number and size distribution of the glaciers of the world. In: Glaciers, ice sheets and volcanoes: a tribute to Mark F. Meier, vol 96, pp 89–94. CRREL Special Report
Messerli B (2000) Mountains of the World—Water Towers for the 21st Century. In This limited edition volume of proceedings from the Rosenberg Forum was prepared and distributed by Mountain Culture at The Banff Centre with funding from the Max Bell Foundation. Permission is hereby granted by the publisher to reproduce this document for non-profit and educational purposes
Nainwal HC, Negi BDS, Chaudhary M, Sajwan KS, Gaurav A (2008) Temporal changes in rate of recession: Evidences from Satopanth and BhagirathKharak glaciers, Uttarakhand, using Total Station Survey. Curr Sci 94(5):653–660
Oerlemans J, Grisogono B (2002) Glacier winds and parameterisation of the related surface heat fluxes. Tellus A 54(5):440–452
Pandey P, Venkataraman G (2013) Changes in the glaciers of Chandra-Bhaga basin, Himachal Himalaya, India, between 1980 and 2010 measured using remote sensing. Int J Remote Sens 34(15):5584–5597
Paul F, Kääb A, Rott H, Shepherd A, Strozzi T, Volden E (2009) GlobGlacier: a new ESA project to map the world's glaciers and ice caps from space. EARSeLeProceedings 8(1):11–25
Qin DH (1999) Map of Glacier Resources in the Himalayas. Science Press, Beijing
Rees HG, Collins DN (2006) Regional differences in response of flow in glacier-fed Himalayan rivers to climatic warming. Hydrol Process 20(10):2157–2169
Racoviteanu A (2011) Himalayan glaciers: combining remote sensing, field techniques and indigenous knowledge to understand spatio-temporal patterns of glacier changes and their impact on water resources. University of Colorado, Boulder, CO. PhD
Racoviteanu AE, Armstrong R, Williams MW (2013) Evaluation of an ice ablation model to estimate the contribution of melting glacier ice to annual discharge in the Nepal Himalaya. Water Resour Res 49(9):5117–5133
Raina VK (2009) Himalayan glaciers: a state-of-art review of glacial studies, glacial retreat and climate change. Himalayan glaciers: a state-of-art review of glacial studies, glacial retreat and climate change. (Report for Min of Environment, Govt of India)
Ren J, Qin D, Kang S, Hou S, Pu J, Jing Z (2004) Glacier variations and climate warming and drying in the central Himalayas. Chin Sci Bull 49(1):65–69
Sarikaya MA, Bishop MP, Shroder JF, Olsenholler JA (2012) Space-based observations of Eastern Hindu Kush glaciers between 1976 and 2007, Afghanistan and Pakistan. Remote Sens Lett 3(1):77–84
Schmidt S, Nüsser M (2012) Changes of high altitude glaciers from 1969 to 2010 in the trans-Himalayan Kang Yatze Massif, Ladakh, northwest India. Arct Antarct Alp Res 44(1): 107–121
Schaner N, Voisin N, Nijssen B, Lettenmaier DP (2012) The contribution of glacier melt to streamflow. Environmental Research Letters 7(3):034029

Scherler D, Bookhagen B, Strecker MR (2011) Spatially variable response of Himalayan glaciers to climate change affected by debris cover. Nature Geosciences 4:156–159
Seidel K, Martinec J, Baumgartner MF (2000) December. Modelling runoff and impact of climate change in large Himalayan basins. In International Conference on Integrated Water Resources Management (ICIWRM), pp 19–21
Sharma AK, Singh SK, Kulkarni AV (2013) Glacier Inventory in Indus, Ganga and Brahmaputra Basins of the Himalaya. Nat Acad Sci Lett 36(5):497–505
Singh P, Arora M, Goel NK (2006) Effect of climate change on runoff of a glacierized Himalayan basin. Hydrol Process 20(9):1979–1992
Singh P, Jain SK (2002) Snow and glacier melt in the Satluj River at Bhakra Dam in the western Himalayan region. Hydrol Sci J 47(1):93–106
Singh RD, Arora M, Kumar R (2010) Snow/Glacier Melt Contribution to Run-off in the Himalayan Rivers-Assessment of Potential Impact of Climate Change
Singh P, Haritashya UK, Kumar N (2008) Modelling and estimation of different components of streamflow for Gangotri Glacier basin, Himalayas/Modélisation et estimation des différentes composantes de l'écoulement fluviatile du bassin du Glacier Gangotri. Himalaya. Hydrological sciences journal 53(2):309–322
Singh P, Jain SK, Kumar N (1997) Estimation of snow and glacier-melt contribution to the Chenab River, Western Himalaya. Mt Res Dev 49–56
Shrestha KL (2005) Global change impact assessment for Himalayan mountain regions for environmental management and sustainable development. Glob Environ Res 9(1):69 (English Edition)
Shrestha AB, Agrawal NK, Alfthan B, Bajracharya SR, Maréchal J, van Oort B (eds) (2015) The Himalayan climate and water Atlas: impact of climate change on water resources in five of Asia's major River Basins. ICIMOD, GRID-Arendal and CICERO
Shi YF (2008) Collection of the studies on glaciology, climate and environmental changes in China. China Meteorological Press, Beijing, pp 850
Shukla A, Gupta RP, Arora MK (2009) Estimation of debris cover and its temporal variation using optical satellite sensor data: a case study in Chenab basin Himalaya. J Glaciology 55(191): 444–452
Srivastava D (2012) Status report on Gangotri glacier. Science and Engineering Research Board, Department of Science and Technology. Himalayan Glaciology Technical Report No. 3, New Delhi, p 10
UNEP Global Environmental Alert Services (2012) Measuring glacier change in the Himalayas. Thematic focus: climate change, ecosystem management. UNEP and WGMS. 2008. Global glacier changes: facts and figures
Vincent C, Ramanathan A, Wagnon P, Dobhal DP, Linda A, Berthier E, Sharma P, Arnaud Y, Azam MF, Gardelle J (2013) Balanced conditions or slight mass gain of glaciers in the Lahaul and Spiti region (Northern India, Himalaya) during the nineties preceded recent mass loss. The Cryosphere 7(2):569–582
Wake C (1989) The influence of summertime precipitation events on the meltwater production in the Karakoram, northern Pakistan. In: Proceedings of the 46th Eastern Snow Conference, pp 28–35
Williams RS Jr., Ferrigno JG (eds) (2012) State of the Earth's cryosphere at the beginning of the 21st century—Glaciers, global snow cover, floating ice, and permafrost and periglacial environments: U.S. Geological Survey Professional Paper 1386–A, 546 p
WGMS (2008) World Glacial Monitoring Service, Glacier Mass Balance Bulletin. 10 (2006–2007), ICSU (FAGS)/ IUGG (IACS)/ UNEP/UNESCO /WMO, Zurich
Xu B, Cao J, Hansen J, Yao T, Joswia DR, Wang N, Wu G, Wang M, Zhao H, Yang W, Liu X (2009) Black soot and the survival of Tibetan glaciers. P Natl Acad Sci USA 106(52):22114–22118

Yao T, Thompson L, Yang W, Yu W, Gao Y, Guo X, Yang X, Duan K, Zhao H, Xu B, Pu J, Lu A, Xiang Y, Kattel DB, Joswiak D (2012) Different glacier status with atmospheric circulations in Tibetan Plateau and surroundings. Nature climate change advance online publication

Zemp M, Hoelzle M, Haeberli W (2009) Six decades of glacier mass-balance observations: a review of the worldwide monitoring network. Ann Glaciol 50(50):101–111

Assessment of Glacier Fluctuations in the Himalaya

Anil V. Kulkarni and S. Pratibha

Abstract Impact of climate change on Himalayan glaciers is debated extensively in India and many scientific data are published in this regard. Yet, there is a gap in our understanding of the current state of glaciers. Hence, we have looked into the scientific studies published on glacier fluctuations in the Himalaya to assess its status. These scientific studies were carried out using both field data and remote sensing techniques. Reconciling of these data suggest overall loss in the glacier area, length and mass across Himalaya, except in the Karakoram region where stable front and advance is observed. On an average, glaciers across the Himalaya were found retreating at a rate of 15.5 ± 11.8 m year^{-1} and have lost an overall area of $13.6 \pm 7.9\%$ from the last four decades. However, these observations on area and retreat are not extensive and are missing from few parts of the Himalaya. Another key parameter to understand health of glacier is mass balance. Mass balance (MB) records are available at glacier, basin and regional scale. Field investigations are carried out at glacier scale and are only available for 14 glaciers which suggest cumulative loss of 20 ± 6 m from 1975 to 2014. However in few glaciers, glaciological measurements do not match very well with geodetic method. Geodetic measurements carried out at basin level show maximum mass loss in Lahaul Spiti region in western Himalaya. On the other hand, region wise monitoring of MB using AAR method suggests maximum loss in the eastern Himalaya and least in the Karakoram. These observations are not free of uncertainty and also show discrepancy between different methods. Therefore, in the wake of rising CO_2 in the atmosphere, extensive field investigations are needed along with development of robust models to understand the glacier response to climate change.

Keywords Accumulation Area Ratio · Mass loss · Retreat

A.V. Kulkarni · S. Pratibha (✉)
Divecha Centre for Climate Change, Indian Institute of Science, Bengaluru, India
e-mail: anilkulkarni@caos.iisc.ernet.in

P.S. Goel et al. (eds.), *Science and Geopolitics of The White World*,
DOI 10.1007/978-3-319-57765-4_13

1 Introduction

Since the industrial revolution, human activities have added large amount of greenhouse gases to the atmosphere. Recent measurements suggest that the current concentration of carbon dioxide in the atmosphere is 400 ppm and is increasing at the rate of 2 ppm $year^{-1}$ (Friedlingstein 2015).

The current rate of increase in the concentration of carbon dioxide due to anthropogenic activity is unprecedented in the earth's history. Resultantly, earth's mean temperature has increased by 0.85 ± 0.2 °C from 1880 to 2012 (IPCC 2014). This can influence glacier distribution and mass balance. Scientific investigations on glacier fluctuations suggest retreat in the Alps, Rockies, Andes and Himalaya. Also, global mass balance records since the beginning of 20th century show the most negative value in last two decades. The global mean mass balance is between −0.54 and −0.80 m w.e $year^{-1}$, depending upon the method used, suggesting that glacier area and mass loss is evident globally (Zemp et al. 2015). However, there are still uncertainty regarding the state and fate of Himalayan glaciers due to the lack of strong scientific evidence and understanding. Recently, a statement was made in the Indian parliament that glacier retreat in the Himalaya shows stable front. This is based on a study of 2018 glaciers in different parts of the Himalaya, where almost 87% of glaciers are found stable (Bahuguna et al. 2014; Times of India 2015). These observations differ from many other published studies (Kulkarni et al. 2007; Bhambri et al. 2012; Kulkarni and Karyakarte 2014; Zemp et al. 2015) Therefore, we have reviewed recent publications to understand the state of glaciers in the Himalaya.

Glacier changes can be quantified as loss in length, area and mass. Glacier retreat and loss in area can be mapped in the field and also, using remote sensing techniques. But, glacier retreat alone may not provide the status of a glacier, as depending upon the glacier characteristic, response time will vary from glacier to glacier. Therefore, the observed retreat will be the integrated reaction to climate change in the past (Bolch et al. 2012). The response time will depend on glacier geometry and MB. In addition, geomorphological features like debris cover and glacial lakes influence retreat. In the Himalaya, large numbers of glaciers are covered in debris (Bhambri et al. 2011). These glaciers retreat differently compared to clean glaciers. A study that compared retreat of two closely located glaciers shows that the clean glacier retreated twice as fast as the debris covered glacier (Lardeux et al. 2015). Further, despite retreating at different rates, the two glaciers have thinned to a similar magnitude, suggesting the limitation of using retreat to infer the state of a glacier. Hence, retreat cannot be used directly as a representative of glacier mass loss. Therefore, mass balance can be a more relevant parameter to understand the influence of climate change. However, they are more challenging to measure than retreat.

2 Change in Length

Change in the glacial length is an indicator of glacier fluctuations. To understand these fluctuations on a larger scale, glacier length changes in different part of Himalaya was compiled from published studies. The data on long-term retreat is so far available for 83 glaciers. The glacier locations are shown in Fig. 1 and details are given in Table 1. The average rate of retreat is 15.5 ± 11.8 m year^{-1} from 1960 to 2014. Mean glacier length of 623 ± 475 m is lost in the past 40 years. The large standard deviation suggests variation in glacier retreat. This variation in retreat is influenced by factors like area altitude distribution, MB, slope, debris cover and glacial lake (Deota et al. 2011; Basnett et al. 2014). We have also combined basin wise retreat for Indus, Ganga and Brahmaputra basins. The amount of retreat at basin wise and of individual glaciers is given in Table 1. Maximum retreat is observed in the Indus basin and least in the Brahmaputra. However, this data set on retreat has more records from western Himalaya, and limited data on retreat of individual glaciers from Eastern Himalaya. Most of the glaciers in the Himalaya are retreating though monitoring of glacier snout in the Karakoram mountain range suggests advance of around 300 m (Bahuguna et al. 2014). However, conclusions based on monitoring of only the snout could be misleading, as retreat is also influenced by factors other than climatic conditions. Therefore, it would be important to monitor other glacier parameters along with long-term changes in glacial extent.

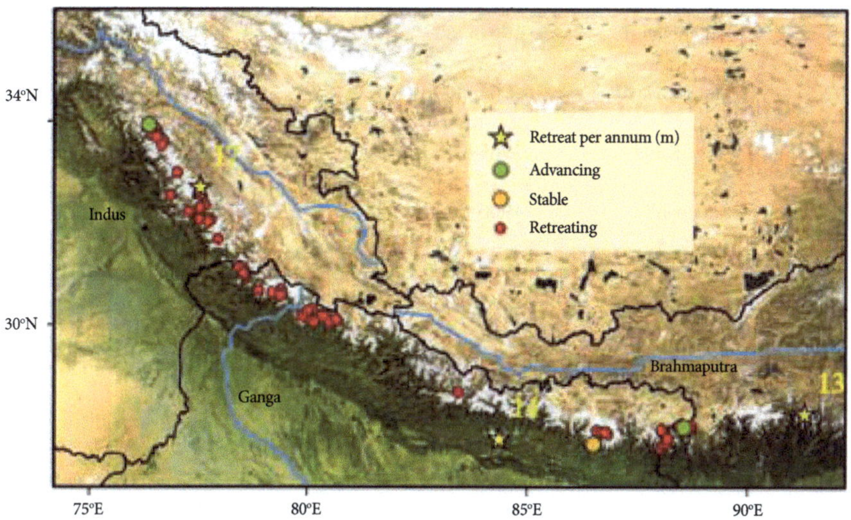

Fig. 1 The location of glaciers with measurements is represented by *circle* and the *colour* of the *circle* represent their condition from past 40 decades, indicating that most of the glaciers are retreating. The basin wise retreat of glaciers for the Indus, Ganga and Brahmaputra is mentioned in *yellow* in m per annum. Further details are given in Table 1 and 2. *Source* Author

Table 1 Retreat of glaciers in the Himalaya

Glacier ID	Glacier	Period	Retreat rate (m/decade)	References
Indus *Average retreat = 178 ± 138 m/decade*				
1	Zanskar glacier 1	1999–2004	80	Kamp et al. (2011)
2	Parkachik	1990–2003	−110	Kamp et al. (2011)
3	Zanskar glacier 3	1990–2003	220	Kamp et al. (2011)
4	Zanskar glacier 4	1975–2003	270	Kamp et al. (2011)
5	Zanskar glacier 5	1990–2003	40	Kamp et al. (2011)
6	Zanskar glacier 6	1990–2003	290	Kamp et al. (2011)
7	Zanskar glacier 8	1975–2003	10	Kamp et al. (2011)
8	Zanskar glacier 7	1975–2003	80	Kamp et al. (2011)
9	Zanskar glacier 9	1975–2006	320	Kamp et al. (2011)
10	Zanskar glacier 10	1975–2006	610	Kamp et al. (2011)
11	Drang Drung	1975–2008	90	Kamp et al. (2011)
12	Zanskar glacier 12	1990–2003	40	Kamp et al. (2011)
13	Zanskar glacier 13	1992–2002	20	Kamp et al. (2011)
14	Zing-Zing Bar (Patsio)	1971–2011	220	Negi et al. (2013)
15	Baralacha	1971–2011	100	Negi et al. (2013)
16	Miyar	1961–1996	160	Sangewar and Kulkarni (2011)
17	Triloknath	1968–1996	180	Sangewar and Kulkarni (2011)
18	Panchinala I	1963–2007	110	Sangewar and Kulkarni (2011)
19	Panchinala II	1963–2007	120	Sangewar and Kulkarni (2011)
20	Beas Kund	1963–2003	190	Sangewar and Kulkarni (2011)
21	Sonapani	1906–1957	180	Sangewar and Kulkarni (2011)
22	Samudra Tapu	1962–2000	200	Kulkarni et al. (2006)
23	Hamtah	1961–2005	80	Sangewar and Kulkarni (2011)
24	Jobri	1963–2003	30	Sangewar and Kulkarni (2011)
25	Chhota Shigri	1962–1995	70	Sangewar and Kulkarni (2011)
26	Sara ugma	1963–2004	410	Sangewar and Kulkarni (2011)
27	Bara Shigri	1906–1995	300	Sangewar and Kulkarni (2011)

(continued)

Table 1 (continued)

Glacier ID	Glacier	Period	Retreat rate (m/decade)	References
28	Man Talai (Gl. No. 115)	1989–2004	230	Sangewar and Kulkarni (2011)
29	Bilare Bange	1962–1997	30	Kulkarni and Bahuguna (2002)
30	Shaune Garang	1962–1997	260	Kulkarni and Bahuguna (2002)
31	Janapa Garang	1962–1997	200	Kulkarni and Bahuguna (2002)
32	Tikku	1960–1999	220	Sangewar and Kulkarni (2011)
33	Jhajju Bamak	1960–1999	280	Sangewar and Kulkarni (2011)
34	Jaundar Bamak	1960–1999	370	Sangewar and Kulkarni (2011)
35	Bandarpunch	1960–1999	260	Sangewar and Kulkarni (2011)
Ganga *Average retreat = 148 ± 97 m/decade*				
36	Dokriani	1962–2007	170	Sangewar and Kulkarni (2011)
37	Gangotri	1935–1996	188	Sangewar and Kulkarni (2011)
37	Gangotri	1962–1999	338	Naithani et al. (2001)
37	Gangotri	1935–2004	220	Kumar et al. (2008)
37	Gangotri	1965–2006	200	Bhambri et al. (2012)
37	Gangotri	1962–2000	397	Bahuguna et al. (2007)
37	Gangotri	2004–2007	119	Sangewar and Kulkarni (2011)
38	Gl. No. 3 (Arwa)	1932–1956	80	Sangewar and Kulkarni (2011)
39	Satopanth	1962–2006	220	Nainwal et al. (2008)
40	Bhagirathi Kharak	1962–2001	167	Sangewar and Kulkarni (2011)
40	Bhagirathi Kharak	1962–2006	73	Nainwal et al. (2008)
41	Trishul bank	1960–2003	220	Sangewar and Kulkarni (2011)
42	Devasthan Bank	1960–2003	260	Sangewar and Kulkarni (2011)
43	Uttari Rishi Bank	1960–2003	340	Sangewar and Kulkarni (2011)

(continued)

Table 1 (continued)

Glacier ID	Glacier	Period	Retreat rate (m/decade)	References
44	Dakshini Rishi Bank	1960–2003	170	Sangewar and Kulkarni (2011)
45	Dakshini Nanda Devi Bank	1960–2003	130	Sangewar and Kulkarni (2011)
46	Milam	1948–1997	170	Sangewar and Kulkarni (2011)
47	Pindari	1906–2001	152	Sangewar and Kulkarni (2011)
47	Pindari	1845–1906	263	de Cotter and Brown (1907)
47	Pindari	1906–1966	184	Tewari (1966)
47	Pindari	1966–2007	64	Bali et al. (2008)
48	Poting	1306–1957	50	Sangewar and Kulkarni (2011)
49	Burphu	1966–1997	50	Sangewar and Kulkarni (2011)
50	Shankalpa	1886–1957	70	Sangewar and Kulkarni (2011)
51	Jhulang	1962–2000	110	Oberoi et al. (2001)
52	Meola	1912–2000	190	Sangewar and Kulkarni (2011)
53	Chipa	1961–2000	270	Oberoi et al. (2001)
54	Nikarchu	1962–2002	90	Sangewar and Kulkarni (2011)
55	Adikailash	1962–2002	130	Sangewar and Kulkarni (2011)
56	Rekha Samba	1974–1999	120	Fujita et al. (2001)
57	AX010	1978–1989	30	Yamada et al. (1992)
58	AX030	1978–1989	0	Yamada et al. (1992)
59	DX 080	1978–1989	50	Yamada et al. (1992)
60	EB 050	1978–1989	30	Yamada et al. (1992)
61	Kongma	1978–1989	30	Yamada et al. (1992)
62	Kongma Tikpe	1978–1989	30	Yamada et al. (1992)
63	Chukhung	1978–1989	90	Yamada et al. (1992)
Brahmaputra *Average retreat = 135 ± 119 m/decade*				
64	Rathong	1976–2005	180	Raina (2009)
65	Onglaklong	1976–2005	100	Raina (2009)
66	Talung	1976–2005	40	Raina (2009)
67	Tongshiong	1976–2005	140	Raina (2009)

(continued)

Table 1 (continued)

Glacier ID	Glacier	Period	Retreat rate (m/decade)	References
68	Zemu	1976–2005	140	Raina (2009)
69	Changsang	1976–2005	220	Raina (2009)
70	E. Langpo	1976–2005	240	Raina (2009)
71	Jongsang	1976–2005	380	Raina (2009)
72	South Lhonak	1962–2008	420	Raj et al. (2013)
73	Lhonak	1976–2005	270	Raina (2009)
74	N. Lhonak	1976–2005	130	Raina (2009)
75	Chuma	1976–2005	80	Raina (2009)
76	Tasha	1976–2005	20	Raina (2009)
77	Tashal	1976–2005	40	Raina (2009)
78	Yulhe	1976–2005	−10	Raina (2009)
79	Changme	1976–2005	30	Raina (2009)
80	Rulak	1976–2005	20	Raina (2009)
81	Tista	1976–2005	150	Raina (2009)
82	Kangkyong	1976–2005	80	Raina (2009)
83	Tenabawa	1976–2005	40	Raina (2009)

3 Change in Areal Extent

In the Himalayas, $\sim$20,060 km^2 out of $\sim$40,000 km^2 area is investigated for the areal loss. It suggests an overall loss of 13.6 $\pm$ 7.9% for 40 years from 1960 onwards. Around 2600 glaciers are studied in the western Himalayan basin and show the highest per cent of area loss, followed by Tista basin. However, only around 325 glaciers are studied in eastern Himalaya. Observed loss is relatively lower in Ganges basin as shown in Fig. 2 and Table 2. Loss in area is observed to be the least in the Karakoram region (Brahmbhatt et al. 2015). In Tista basin, loss is much higher than in Ganga and Upper Indus basin even though glaciers are highly debris covered. Recent investigation has shown that excessive debris can lead to formation of supraglacial lakes leading to a faster areal loss (Basnett et al. 2014). This investigation has shown that the loss in area due to lake formation will be 5 $\pm$ 1.3% higher than only debris covered glaciers. Therefore, glaciers in the other parts of Himalaya also are susceptible to this phenomenon and it needs continuous monitoring of Himalayan glaciers.

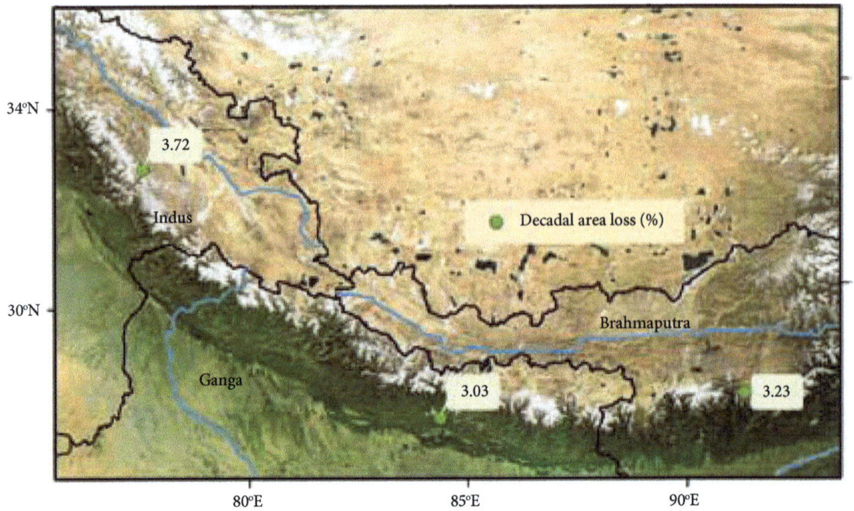

Fig. 2 Basin wise glacial area loss (%) per decade since last 40 years in the Himalaya. *Source* Author

Table 2 Basin wise area loss in the Himalaya

Basin	Percent area loss/decade	Standard deviation
Indus	3.72	1.93
Ganga	3.03	2.26
Brahmaputra	3.22	1.83

4 Mass Loss

Mass balance (MB) investigation helps in monitoring fluctuations in glacier mass and analyse the state of a glacier better than retreat and area loss. Mass balance is also useful to model future changes in glacial extents and river run off pattern (Kulkarni et al. 2004). Traditionally, glaciological method is adopted that involves measuring winter snow accumulation using pits and summer ice melt using stakes in ablation area. This requires extensive field work on glacier and is very challenging in remote and rugged Himalayan Terrain. Hence, mass balance records from field are available for a limited number of glaciers and for short periods.

So far, MB records are available for 11 glaciers in western, four in Central and one in eastern Himalaya (Pratap et al. 2015). The first MB measurement for an Indian glacier was carried out in 1974 in Gara glacier, western Himalaya, where positive Mass budget was observed in 1975 and 1976 (Raina et al. 1977). Besides this, Gor Garang and Shaune Garang, also located in the western Himalaya, exhibited positive MB in 1983. Also, Nehnar in western Himalaya, Tipra in Central and Changmekangpu in eastern Himalaya showed mass gain in the same year.

However, MB records in 90s show negative trend. Negative MB of −0.83 m w.e. was observed in Shaune Garang in 1991. A Consistent negative MB was observed in Hamtah glacier from 2000 to 2012 and interestingly, unlike the nearby glacier Chhota Shigri which experienced stable mass balance in 2004/2005 and 2008–2011 (Pratap et al. 2015; Wagnon et al. 2007; Azam et al. 2012). Overall, cumulative specific mass balance of Indian glaciers suggests mass loss. Field MB data of 14 glaciers suggests that the mean mass loss from 1975 to 2014 is 20 ± 6 m as shown in Fig. 3. The name and location of these glaciers are shown in Fig. 4.

However, field measurements are not devoid of limitations. Measurements are done at specific points on the glacier and extrapolation of point measurements to entire glacier induces uncertainty. In addition, avalanche accumulation can also introduce uncertainties in estimates of winter mass gain. In situ measurements tend to be biased towards small to medium sized debris free glaciers, as they are easy to access. Therefore, to get mass balance on larger scale, we need to consider alternate methods such as Accumulation Area Ratio or Geodetic.

In Geodetic method, change in elevation at two different periods is computed to determine the mass balance. Geodetic MB measurements have been carried out at glacier and at regional level (Pratap et al. 2015). Glaciological MB data of Chhota Shigri glacier from 2002 was extended back to 1988 using geodetic method using field and space borne data. Though a slight mass gain or a balanced mass budget

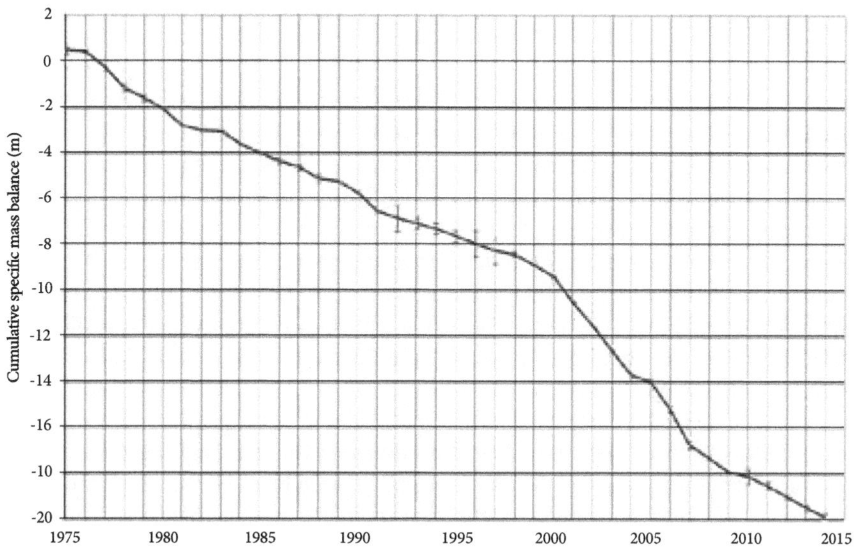

Fig. 3 Cumulative glacial mass loss from 1975 to 2014 for 14 glaciers is estimated at 20 ± 6 m, where uncertainly represent cumulative error. The *error bars* in the graph represent the uncertainty in mass loss estimation for individual year. Predominantly, field mass balance data was used to estimate the cumulative mass balance. However, for few glaciers and years, gaps in mass balance data were filled using ELA and AAR methods. The cumulative mass balance will change, if data from other regions such as Sikkim and Karakoram are available. *Source* Author

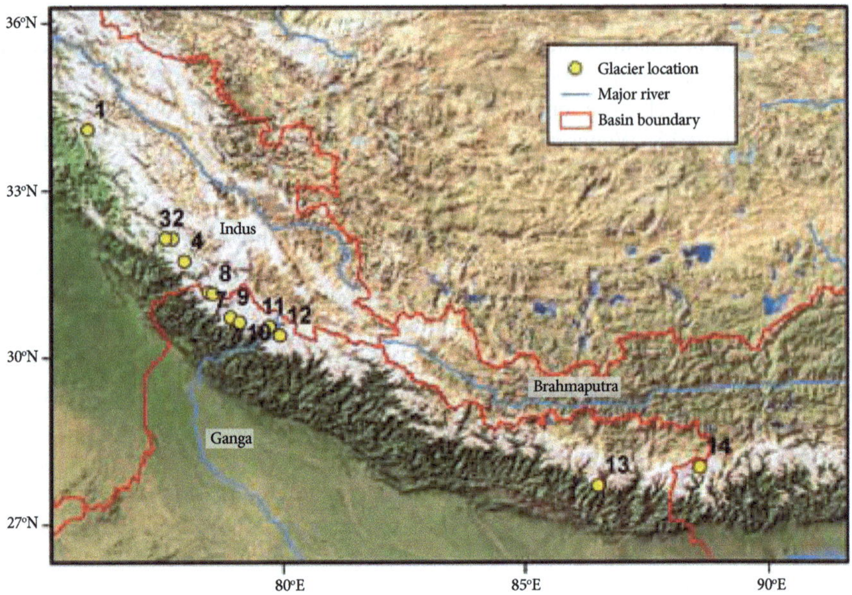

Fig. 4 Location of 14 glaciers, where field based estimates of mass balance are available, *1* Nehnar, *2* Hamta, *3* Chhota Shigri, *4* Parbati, *5* Gara, *6* Gor Garang, *7* Shaune Garang, *8* Naradu, *9* Dokriani, *10* Chorabari, *11* Tipra bank, *12* Dunagiri, *13* AX010 and *14* Changme Khangpu. *Source* Author

was observed between 1988 and 1999, mass loss was observed post 1999 and overall, mass loss of –0.17 ± 0.08 m w.e. year^{-1} was observed from 1988 to 2011, whereas, geodetic measurements in Hamtah glacier show strongly negative MB of –0.45 ± 0.16 m w.e. year^{-1} from 1999 to 2011. However, glaciological measurements suggest higher mass loss of –1.46 m w.e. year^{-1} from 2000 to 2009 and hence, does not match with geodetic measurements (Vincent et al. 2013). At regional level, Geodetic method using space borne data was applied in Lahaul–Spiti, Western Himalaya, from 1999–2004 which suggests overall specific mass balance of −0.7 to −0.85 m w.e. year^{-1} (Berthier et al. 2007). MB data of Nepal Himalaya also suggests negative mass budget of −0.32 ± 0.13 m w.e. year^{-1} from 1999 to 2011. On the other hand, slight mass gain of +0.11 ± 0.14 m w.e. year^{-1} and +0.09 ± 0.18 m w.e. year^{-1} was observed in the east and west Karakoram region, respectively, during the same period (Gardelle et al. 2013). Overall, the mass changes observed using geodetic measurements are heterogeneous across the Himalaya. Slight mass gain was observed in Karakoram region and moderate loss observed in eastern Himalaya. However, maximum mass loss was observed in Spiti Lahaul (western Himalaya). Use of remotely sensed data makes geodetic method efficient in observing mass changes over longer periods and larger spatial area. It can also provide historical mass changes if DEM is available. However, mass loss estimated from elevation changes are not free of biases and need addressed through various corrections (Gardelle et al. 2013).

However, to understand future changes in mass balance, we need alternate methods such as Accumulation Area Ratio, where ELA can be modelled using temperature and precipitation. Conventionally, Empirical Relationship between the specific mass balance and AAR is used to estimate mass balance. This was applied at Gara and Gor-Gorang glacier, Himachal Pradesh, by Kulkarni (1992). This method allows monitoring of mass changes of large number of glaciers, where ELA can be delineated using satellite imagery. Further, this method was applied to 19 glaciers in Western Himalaya and specific mass balance was estimated as −90 and −78 cm, for 2000 and 2001 respectively. Estimate of mass changes in 32 glaciers in Tirungkhad river basin, western Himalayan by AAR (Area Accumulation Ratio) method shows positive mean mass balance for 5 glaciers but overall specific mass balance was observed to be −27 cm (2000) to −41 cm (2011). AAR method was applied over the entire Himalayan region by Chaturvedi et al. (2014). The specific mass balance for the year 2000 for Karakoram region, Western Himalaya, *Central Himalaya and Eastern Himalaya is estimated at* +171 ± 63, −165 ± 80, −508 ± 69, *and* −809 ± 38 mm w.e.year^{-1} *respectively and* −162 ± 16 mm w.e. year^{-1} for the entire Karakoram-Himalayan region. However, in contrast to MB estimates by geodetic method, maximum mass loss is estimated in eastern Himalaya by AAR method. This technique is also not free of errors, as position of snow line at the end of summer is difficult to obtain at times. Therefore, further improvement in methodology is carried out, where data such as transient snowline, temperature and precipitation is used (Tawde et al. 2016). As this methodology uses temperature and precipitation, future changes in these parameters can be obtained using climate models. This will help us in predicting future changes in glacier mass loss and subsequent glacier extent.

References

Azam MF, Wagnon P, Ramanathan A, Vincent C, Sharma P, Arnaud Y, Linda A, Pottakkal JG, Chevallier P, Singh VB, Berthier E (2012) From balance to imbalance: a shift in the dynamic behavior of Chhota Shigri Glacier (Western Himalaya, India). J Glaciol 58(208):315–324

Bahuguna IM, Kulkarni AV, Nayak S, Rathore BP, Negi HS, Mathur P (2007) Himalayan glacier retreat using IRS 1C PAN stereo data. Int J Remote Sens 28(2):437–442

Bahuguna IM, Rathore BP, Brahmbhatt R, Sharma M, Dhar S, Randhawa SS, Kumar K, Romshoo S,. Shah RD, Ganjoo RK, Ajai (2014) Are the Himalayan glaciers retreating? Curr Sci 106(7)

Bali R, Agarwal KK, Ali N, Rastogi SK (2008) Geomorphological framework and recessional behavior of the Pindari glacier, Kumauon Himalaya. In: National Seminar on Glacial Geomorphology and Paleoglaciation in Himalaya, Lucknow, pp 32–33

Basnett S, Kulkarni AV, Bolch T (2014) The influence of debris cover and glacial lakes on the recession of glaciers in Sikkim Himalaya, India. J Glaciol 59(218):2013. doi:10.3189/2013JoG12J184

Berthier E, Arnaud Y, Kumar R, Ahmad S, Wagnon P, Chevallier P (2007) Remote sensing estimates of glacier mass balances in the Himachal Pradesh (Western Himalaya, India). Remote Sens Environ 108(3):327–338. doi:10.1016/j.rse.2006.11.017

Bhambri R, Bolch T, Chaujar RK (2011) Mapping of debris covered glaciers in the Garhwal Himalayas using ASTER DEMs and thermal data. Int J Remote Sens 32(23):8095–8119. doi:10.1080/01431161.2010.532821

Bhambri R, Bolch T, Chaujar RK (2012) Frontal recession of Gangotri Glacier, Garhwal Himalayas, from 1965 to 2006, measured through high-resolution remote sensing data. Curr Sci 102(3):489–494

Bolch T, Kulkarni A, Kääb C, Huggel F, Paul JG, Cogley Frey H, Kargel JS, Fujita K, Scheel M, Bajracharya S, Stoffel M (2012) The state and fate of Himalayan glaciers. Science 336 (6079):310–314. doi:10.1126/science.1215828

Brahmbhatt RM, Bahuguna IM, Rathore BP, Singh SK, Rajawat AS, Shah RD, Kargel JS (2015) Satellite monitoring of glaciers in the Karakoram from 1977 to 2013: an overall almost stable population of dynamic glaciers. Cryosphere Discuss 9:1555–1592

de Cotter GP, Brown J (1907) Notes on certain glaciers in Kumaun. Rec Geol Surv India 35(4), 148–157

Chaturvedi RK, Kulkarni Anil, Karyakarte Yogesh, Joshi J, Bala G (2014) Glacial mass balance changes in the Karakoram and Himalaya based on CMIP5 multi-model climate projections. Clim Change 123:315–328

Deota BS, Trivedi YN, Kulkarni AV, Bahuguna IM, Rathore BP (2011) RS and GIS in mapping of geomorphic records and understanding the local controls of glacial retreat from the Baspa Valley, Himachal Pradesh, India. Curr Sci 100(10):1555–1563

Fujita K, Nakazawa F, Rana B (2001) Glaciological observations on Rikha Samba glacier in Hidden Valley, Nepal Himalayas, 1998 and 1999. Bull Glacier Res 18:31–35

Friedlingstein P (2015) Carbon cycle feedbacks and future climate change. Philos Trans R Soc A 373:20140421. doi:10.1098/rsta.2014.0421

Gardelle J, Berthier E, Arnaud Y, Kaab A 2013. Region-wide glacier mass balances over the Pamir–Karakoram–Himalaya during 1999–2011. Cryosphere 7:1263–1286. doi:10.5194/tc-7-1263-2013

IPCC (2014) Climate change 2014: synthesis report. Contribution of working groups I, II and III to the fifth assessment report of the Intergovernmental Panel on Climate Change. In: Core Writing Team, Pachauri RK, Meyer LA (eds). IPCC, Geneva, Switzerland, 151 pp

Kamp U, Byrne M, Bolch T (2011) Glacier fluctuations between 1975 and 2008 in the Greater Himalayan Range of Zanskar, Southern Ladakh. J Mt Sci 8:374–389

Kulkarni AV (1992) Mass balance of Himalayan glaciers using AAR and ELA methods. J Glaciol 38:101–104

Kulkarni AV, Bahuguna IM (2002) Glacial retreat in the Baspa Basin, Himalayas, monitored with satellite stereo data. J Glaciol 48(160):171–172

Kulkarni AV, Dhar S, Rathore BP, Raj KBG, Kalia R (2006) Recession of Samundra Tapu glacier, Chadra river basin, Himachal Pradesh. J Indian Soc Remote Sens 34(1):39–46

Kulkarni AV et al (2007) Glacial retreat in Himalaya using Indian remote sensing satellite data. Curr Sci 92(1):69–74

Kumar K, Dumka RK, Miral MS, Satyal GS, Pant M (2008) Estimation of retreat rate of Gangotri glacier using rapid static and kinematic GPS survey. Curr Sci 92(2):258–262

Kulkarni AV, Karyakarte Y (2014) Observed changes in Himalayan Glaciers. Curr Sci 106 (02):237–244

Lardeux P, Glasser N, Holt T, Fynn TI, Hubbard B (2015) Area and elevation changes of a Debris-Covered glacier and a Clean-Ice Glacier between 1952–2013 using Aerial Images and Structure-from-Motion. Abstract submitted at AGU, Fall meeting, San Francisco

Naithani AK, Nainwal HC, Sati KK, Prasad C (2001) Geomorphological evidences of retreat of the Gangotri glacier and its characteristics. Curr Sci 80(1):87–94

Nainwal HC, Negi BDS, Chaudhary M, Sajwan KS, Gaurav A (2008) Temporal changes in rate of recession: evidences from Satopanth and Bhagirath Kharak glaciers, Uttarakhand, using Total Station Survey. Curr Sci 94(5):653–660

Negi HS, Saravana G, Rout R, Snehmani (2013) Monitoring of great Himalayan glaciers in Patsio region, India using remote sensing and climatic observations. Curr Sci 105(10):1383–1392 (25 Nov 2013)

Oberoi LK, Siddiqui MA, Srivastava D (2001) Recession of Chipa, Meola and Jhulang (Kharsa) glaciers in Dhauliganga Valley between 1912 and 2000. GSI Special Publication No. 65, 2001, vol II, pp 57–60

Pratap B, Dobhal DP, Bhambri R, Mehta M, Tewari CV (2015) Four decades of glacier mass balance observations in the Indian Himalaya. Reg Environ Change 16:643. doi:10.1007/s10113-015-0791-4

Raina VK, Kaul MK, Singh S (1977) Mass balance studies of Gara glacier. J Glaciol 18(80):415–423

Raina VK (2009) Himalayan glaciers: a state-of-art review of glacial studies, glacial retreat and climate change. MoEF Discussion Paper 56

Raj KBG, Remya SN, Kumar KV (2013) Remote sensing based hazard assessment of glacial lakes in Sikkim Himalaya. Curr Sci 104(3):359–364

Sangewar CV, Kulkarni AV (2011) Observational studies of the recent past. In: Report of the Study Group on Himalayan glaciers prepared for the Office of the Principal Scientific Adviser to the Government of India, PSA/2011/2, pp 25–76

Tawde SA, Kulkarni AV, Bala G (2016) Estimation of glacier mass balanceon a basin scale: an approach based on satellite-derived snowlines and a temperature index model. Curr Sci 111 (12):1977–1989

Tewari AP (1966) Recent changes in the position of the snout of the Pindari glacier (Kumaun Himalaya) Almora District, Uttar Pradesh, India. In International Symposia on the Role of Snow and Ice in Hydrology, International Association of Hydrological Science, vol 106, pp 1144–1149

Times of India (2015) 87% of Himalayan glaciers stable since 2001: Javadekar (Dec 15)

Vincent C, Ramanathan AL, Wagnon P, Dobhal DP, Linda A, Berthier E, Sharma P, Arnaud Y, Azam MF, Jose PG, Gardelle J (2013) Balanced conditions or slight mass gain of glaciers in the Lahaul and Spiti region (northern India, Himalaya) during the nineties preceded recent mass loss. Cryosphere 7:569–582. doi:10.5194/tc-7-569-2013

Wagnon P, Linda A, Arnaud Y, Kumar R, Sharma P, Vincent C, Pottakal JG, Berthier E, Ramanathan A, Hasnain SI, Chevallier P (2007) Four years of mass balance on Chhota Shigri Glacier, Himachal Pradesh, India, a new benchmark glacier in the western Himalaya. J Glaciol 53(183):603–611

Yamada T et al (1992) Fluctuations of the Glaciers from 1970s to 1989 in the Khumbu, Shorong and Langtang region, Nepal Himalayas. Bull Glacier Res 10:11–19

Zemp M et al (2015) Historically unprecedented global glacier decline in the early 21st century. J Glaciol 61(228):745–763. doi:10.3189/2015JoG15J017